D1161940

Burley

BURLEY

Kentucky Tobacco in a New Century

ANN K. FERRELL

Scholarly publisher for the Commonwealth,
serving Bellarmine University, Berea College, Centre College of Kentucky, Eastern Kentucky University, The Filson Historical Society, Georgetown College, Kentucky Historical Society, Kentucky State University, Morehead State University, Murray State University, Northern Kentucky University, Transylvania University, University of Kentucky, University of Louisville, and Western Kentucky University.

Editorial and Sales Offices: The University Press of Kentucky
663 South Limestone Street, Lexington, Kentucky 40508-4008
www.kentuckypress.com

17 16 15 14 13 5 4 3 2 1

Library of Congress Cataloging-in-Publication Data

Ferrell, Ann K., 1972-
Burley : Kentucky tobacco in a new century / Ann K. Ferrell.
pages cm. — (Kentucky remembered : an oral history series)
Includes bibliographical references and index.
ISBN 978-0-8131-4233-3 (hardcover : acid-free paper) —
ISBN 978-0-8131-4234-0 (epub) — ISBN 978-0-8131-4235-7 (pdf)
1. Burley tobacco—Kentucky. 2. Burley tobacco—Kentucky—History. 3. Tobacco farms—Kentucky. 4. Tobacco farmers—Kentucky—Biography. 5. Oral history—Kentucky. 6. Tobacco—Social aspects—Kentucky. 7. Social change—Kentucky. 8. Kentucky—Social life and customs. 9. Kentucky—Economic conditions. I. Title.
SB273.F47 2013
633.7'109769—dc23 2013001705

This book is printed on acid-free paper meeting the requirements of the American National Standard for Permanence in Paper for Printed Library Materials.

Manufactured in the United States of America.

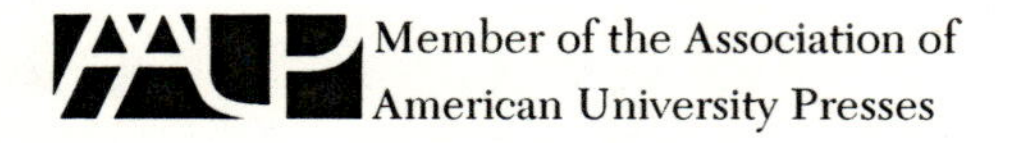

Contents

Illustrations

Maps

Tables

Figures

Photographs

Series Foreword

In the field of oral history, Kentucky is a national leader. Over the past several decades, thousands of its citizens have been interviewed. While oral history is, of course, only one type of source material, the very personal nature of recollection often discloses hidden aspects of history. Oral sources thus provide a vital thread in the rich fabric that is Kentucky history. Kentucky Remembered: An Oral History Series brings into print the most important of the state's collections of its citizens' oral history, with each volume focusing on a particular subject.

Burley: Kentucky Tobacco in a New Century is the twelfth volume in the Kentucky Remembered series, which explores the ways in which oral history connects the individual story to tradition and to the historical record. Kentucky's transitioning family farms were extensively documented in the 1990s for the Kentucky Family Farm Oral History Project. Over five hundred oral history interviews were recorded and archived, primarily documenting major agricultural, economic, and technological transitions. Ann K. Ferrell has continued this tradition of documentation, using oral history to explore the history and culture of tobacco farming as it has transitioned and declined on the Kentucky landscape. Her book *Burley: Kentucky Tobacco in a New Century* takes the stories of individual farmers and uses them as a lens to closely examine the meanings underlying the region's complex relationship with tobacco, providing a greater understanding of the farmers whose lives, family stories, and cultures are inextricably linked to this controversial crop.

James C. Klotter
Terry L. Birdwhistell
Douglas A. Boyd

A Note on Transcription

The process of transcribing the spoken word onto the page and presenting it for others to read is a harrowing one; none of us speaks in the complete, grammatically correct sentences that we might hope. Like others who face this challenge, I have struggled with how to represent the words of my interviewees in a manner that is respectful, readable, and true to their voices. In this spirit, I have made minor edits to the verbatim transcriptions in order to reflect the translation of the spoken word onto the page. This means that I have deleted utterances that are not germane to a speaker's meaning, including utterances such as *uh* and *um,* as well as repeated words and phrases (unless they are used for emphasis or some other specific meaning) and false starts. However, I have not followed standard rules of punctuation in most cases and instead use punctuation to reflect the cadence of talk. This means, for instance, that if a speaker clearly uttered an incomplete sentence (as speakers so often do) in order to emphasize something he or she said, I have left that sentence incomplete. In other instances, commas are used to denote brief pauses. The use of ellipses (. . .) indicates long pauses, and the use of an em-dash (—) indicates the speaker stopped in midsentence and redirected her- or himself toward a new thought. I only changed whole words in cases in which I replaced nonstandard verb-tense usage with standard usage (e.g., *was* to *were*). Leaving off the g at the end of a verb (*goin'*) is common in spoken language, but here I have filled these letters in because such spellings tend to be a distraction to the reader. However, I have left words such as *gonna* (rather than *going to*) because although they are nonstandard, they are commonly used. Occasionally, I have added a word or phrase that was evident in the context of the interview but was not said directly, and in all such instances I have used [brackets] to denote additions.

Acknowledgments

No project can be completed without the input, support, and assistance of many individuals and institutions. This is especially true in the case of one based on ethnographic fieldwork with those who are the experts on the topic. Most of all, I am grateful to the farmers and farm family members, tobacco warehousemen, University of Kentucky College of Agriculture and Kentucky Cooperative Extension Service personnel, and many others who so generously and graciously shared their time and knowledge with me. All those I interviewed are listed in the bibliography, but I would particularly like to thank those who allowed me to visit with them repeatedly: Alice Baesler, Keenan Bishop, Jerry and Kathleen Bond, Mike Carter, Ben Crain, George Duncan, Clarence Gallagher, Martin and Kathy Henson, Lisa and Roger Perkins, Wilbert and Patsy Perkins, Roger Quarles, Jerry Rankin, G. B. and Jonathan Shell, and Marlon Waits. Not all those I learned from are cited by name in this book, but the knowledge I gained from every person I met over the course of this project informs this work.

I also owe a great debt to the many individuals who introduced me to farmers and others in their communities who are knowledgeable about tobacco, including Keenan Bishop, Edwin Chavous, George Duncan, Kara Keeton, Steve Moore, Carol Shutt, Charlene Smith, and Diana Taylor. I am extremely grateful to the many people who read earlier (some much earlier) versions of parts or all of this manuscript. This includes, first and foremost, Amy Shuman, Dorothy Noyes, Patrick B. Mullen, and Nan Johnson, as well as Brent Björkman, Erika Brady, George Duncan, Martin Henson, Robert Pearce, Martha Sims, and the anonymous readers who read it on behalf of the University Press of Kentucky. This project grew, evolved, and became stronger because of the suggestions, corrections, and feedback provided by all of these people. I also wish to thank Will Snell of the University of Kentucky College of Agriculture for answering data-related questions at many points over the course of this project and Bob Gray for assistance with the graphs included here. Papers related to this work were presented at annual meetings of the American Folklore Society and conferences sponsored by the Kentucky Oral History Commission and the Ohio State University Center for Folklore Studies over the years, and I am grateful for the comments from and conversations with those in attendance and those who served on panels with me, including Sheila Bock, Larry Danielson, Diane Goldstein, Patrick B. Mullen, Amy Shuman, C. W. Sullivan III, and Jason Whitesel. Thanks also to Katherine Chappell and Savannah Napier for assistance with bibliographic citations, to Afsaneh Rezaeisahraei for assistance in final preparations for publication, to Carol Sickman-Garner for her keen eye on the final manuscript, and to Anne Dean Watkins and Bailey Johnson at the University Press of Kentucky for their commitment to seeing this book through to completion. During a conversation about my research, Henry Glassie suggested that the most fitting title for this book is simply *Burley;* I thank him for that. At the same time that I express gratitude to all of these people I, of course, accept the blame for any inaccuracies.

I was able to complete this project because of the ongoing support of family and friends. All of my parents and my extended family have encouraged me throughout the process. My fieldwork was made possible by the hospitality, generosity, and friendship of Diana Taylor and Bob Gray. Throughout this journey, I have been accompanied (if not always physically) by my best friend and partner in life and folklore, Brent Björkman. Thank you, Brent, for believing that I could do this and for all that you have done to help make it possible.

I wish to thank the following organizations and institutions for financial support that made my research, writing, and publication possible: the Kentucky Oral History Commission, Kentucky Historical Society; the Ohio State University Center for Folklore Studies, College of the Humanities, and Department of English; the American Association of University Women; and Potter College of Arts and Letters, Western Kentucky University. I have benefited from the support of my colleagues in the Department of Folk Studies and Anthropology at Western Kentucky University, headed by Michael Ann Williams. I am grateful for the staff and resources of the Kentucky Department for Libraries and Archives and the libraries of the Ohio State University, Middlebury College, and Western Kentucky University. The digital recordings of all interviews that I conducted over the course of my research are deposited in the archives of the Kentucky Oral History Commission (KOHC), Kentucky Historical Society, Frankfort. This project also benefited greatly from interviews conducted by John Klee, Lynne David, and Kara Keeton for the KOHC. A list of all of their interviews that I cite directly or that I was informed by can also be found at the end of this book. All interviews archived with the KOHC are cited here with permission.

Introduction

"Would you rather have present day or olden days?"

Tradition and Transition in Kentucky Burley Tobacco Production

Frequently over the past decade, I have heard Kentucky natives comment with sadness on the changing landscape of their home state: the countryside of childhood will soon be gone. The links between land and culture, sense of place, history, and identity have been widely acknowledged.[1] According to Lucy Lippard, "The intersections of nature, culture, history, and ideology form the ground on which we stand—our land, our place, the local."[2] Such intersections, of course, are neither inherent in the land itself nor static. *We* form the ground on which we stand through our use of it and as we come to view it not just as land but as landscape. Gregory Clark argues, "*Land* becomes *landscape* when it is assigned the role of symbol, and as symbol it functions rhetorically."[3] It is precisely because landscapes symbolize something about who we are that shifting landscapes often result in feelings of loss. The source of the sense of loss expressed by so many Kentuckians is not the expected—the loss of land to the proliferation of subdivisions and "big-box" retail stores, although certainly many bemoan such development. This sense of loss follows observations that the tobacco fields are disappearing.

It is difficult for many to understand the loss of tobacco—a crop that has come to symbolize addiction, disease, and a deceptive industry—as lamentable. However, this loss has vast economic and cultural consequences for farming communities, as well as for the state as a whole. Tobacco was once Kentucky's largest cash crop, and although other types of tobacco are grown in the state, historically over 90 percent of the tobacco grown in Kentucky has been burley tobacco.[4] The crop has been an important symbol of regional identity, and the changing landscape symbolizes a shift to a "brave new world"[5] in which King Burley no longer reigns and the future is uncertain—both economically and symbolically. I came to this project thinking that it would somehow be possible to conduct research with tobacco farmers disconnected from tobacco products. I found that not only is that not possible but that the interconnections are central to the story. This book is premised on the idea that the stories of tobacco farmers and of burley tobacco in Kentucky must be understood not separate from but within the context of the changing meanings of the crop.

In the interwoven process of collecting and interpreting the material that forms the basis of this book, I bring together my training as a folklorist with the theory and methods of the field of rhetoric, particularly the work of Kenneth Burke.[6] In bringing the two together, not only do I view the performance of traditional cultural practices as persuasive and attempt to understand the rhetorical force of the usage of terms such as *tradition* and *heritage,* but I also investigate the interactions between the performance of cultural practices and public discourses about such practices. "Public discourses" have been examined in a range of fields, often involving questions of what constitutes public and how various media produce, sustain, change, or limit understanding of an issue. In his rhetorical reading of American tourist landscapes, Gregory Clark defines "public discourse" as "the ongoing process of inquiry and exchange that is sustained by people who constitute . . . community."[7] Clark understands public discourses not only as "tak[ing] the form of print and speech" but also as "experiences not immediately discursive at all."[8] My interest is in the emerging and evolving discourses surrounding tobacco farming in the context of other public discourses on tobacco, such as those related to smoking, health, and disease, as well as those related to farming more generally, such as increasing calls in recent years for the procurement and consumption of foods grown by local farmers. This requires an understanding of the historical conditions in which these multiple discourses have emerged and the ways in which they compete.

It also requires an understanding of the discourses of tobacco farmers.

This book is the result of the collection of two kinds of data, broadly speaking. First, I rely on data that I gathered through ethnographic fieldwork primarily in central and northern Kentucky, the center of burley tobacco production. Research began on this project in 2005, with intensive fieldwork during the 2007 crop year—January 2007 through February 2008. During this period, I spent time on farms, observing and at times participating in tobacco production. I conducted one or more recorded interviews with over sixty farmers, warehousemen, and agricultural professionals, and I visited the farms of and had conversations with many others. In addition to my own interviews, with the support of a grant from the Kentucky Oral History Commission (KOHC), I fully transcribed thirty-three recorded interviews conducted by John Klee and Lynne David for the KOHC between May 2000 and February 2002 with farmers, agriculture professionals, and policy makers. I attended agricultural events such as trainings, meetings, and field days, and I visited public and private sites related to tobacco once it leaves the farm, such as one of the last remaining burley tobacco warehouses, tobacco receiving stations (where farmers now sell their crop), and a redrying facility (where tobacco is processed before it is shipped to manufacturing facilities). I also worked in the tobacco exhibit area of the Kentucky Folklife Festival in 2005 and 2007, interpreting tobacco traditions for visitors alongside farmers, interacting with and observing visitors to the tobacco tent, and conducting interviews both on and off stage.

Second, throughout the period of my research I have observed and collected (both systematically and serendipitously) public discourses from a wide range of oral, print, and electronic sources. I spent many days in the Kentucky Department for Libraries and Archives, reading the newsletters of the Kentucky Department of Agriculture, and I collected published materials—books, pamphlets, brochures, booklets, fliers, posters, and policy and statistical documents—about tobacco history and production from a range of sources, including tobacco industry lobbying and marketing organizations, tobacco companies, the state and federal government, research institutions, and farm organizations. During my time in the field, I read the major Kentucky newspapers on a daily basis, I utilized online databases to locate media coverage in past

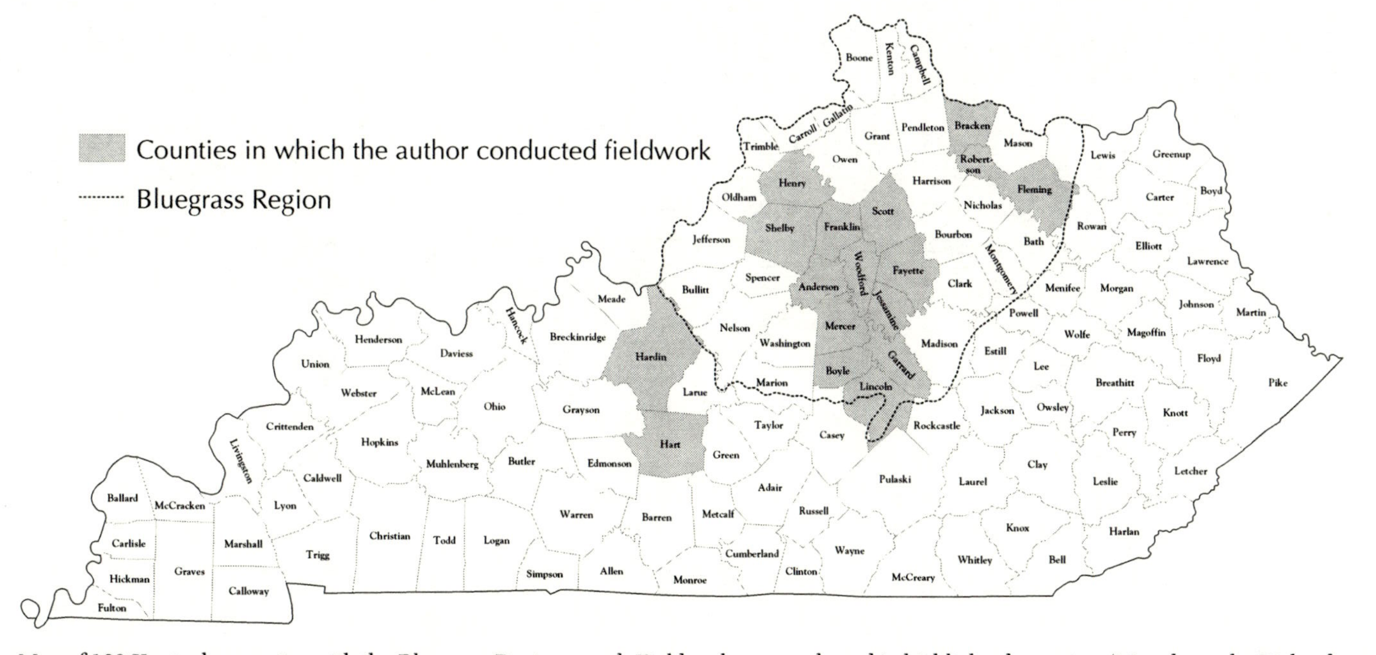

Map of 120 Kentucky counties with the Bluegrass Region noted. Fieldwork was conducted in highlighted counties. (Map drawn by Richard A. Gilbreath)

periods, and I regularly visited and studied a number of websites, particularly those of state-sponsored agriculture agencies and farm organizations.

In a caveat meant to explain how his book *The Written Suburb: An American Site, An Ethnographic Dilemma* departs from "conventions of 'traditional' ethnographic documentation," John Dorst describes such documentation as "an attempt to 'get close to,' to understand and to describe the culture, values or world view of a certain set of people."[9] This was the task of my ethnographic fieldwork with tobacco farmers; I set out not with a hypothesis to prove, but with a wish to understand not only tobacco farming practices but the meanings of these practices to farmers during the current period of transition.[10] In part, I have gathered a retrospective history, an accounting of how people today view the past and their connections to it. I quickly understood that these meanings can only be understood within a context of the public discourses that surround these farmers and their ongoing processes of the generation of meaning(s). As they talked to me, they were also talking back to those who argue that tobacco is being replaced by diversified farming, to the perception of tobacco as a thing of the past (as "heritage"), and of course to those who argue that the crop that they depend upon is lethal. These and other discourses surrounding tobacco serve as screens through which tobacco is differently understood. Kenneth Burke writes: "When I speak of 'terministic screens,' I have particularly in mind some photographs I once saw. They were *different* photographs of the *same* objects, the difference being that they were made with different color filters. Here something so 'factual' as a photograph revealed notable distinctions in texture, and even in form, depending on which color filter was used for the documentary description of the event being recorded."[11] Tobacco farming (and more important, tobacco farmers) takes on differing, often competing, textures and meanings dependent on the discursive screen through which it is considered. My folklore training helps me to understand how people communicate (with each other, with me) on the ground; rhetoric helps me to understand the persuasive work of discourses (both on the ground and seemingly all around, in more public discourses) and what is included and left out in order for persuasion to take place. Together, they help me to understand interactions between multiple domains—rather than attempting to separate public from private or cultural practices from how they are represented in multiple sites.

Of course, my endeavor to understand tobacco farmers within the context of such public discourses and the changing symbolism of the crop is just that: *my* endeavor. In recent decades, ethnographers in the fields of folklore, anthropology, and others have problematized prior assumptions of the writing of ethnography as an objective enterprise.[12] Rather, as ethnographers, we offer interpretations that are filtered through many screens of our own, both scholarly and personal. I also came to my fieldwork as an outsider, and I write as an outsider; this both limits my ability to understand tobacco culture from an insider perspective and affords me the opportunity to offer an interpretation from the outside looking in. My assemblage and use of sources from multiple domains—ethnographic, historical, archival—engenders a dialogic account rather than one that relies on any one voice.

This book must also be read as specific to the people with whom I interacted and the region in which they live and work. Geography and climate led to the establishment of distinct tobacco regions based on the type of tobacco grown. Cultivation methods, labor, marketing practices, and farm size all vary by region. The historical context of tobacco production is unique to each tobacco region as well, because tobacco type, farm size, and landscape determined, for instance, the degree of dependence on slave labor and, later, the mechanization of tobacco work.[13] Central Kentucky has historically been the center of burley tobacco production, the center of what is referred to as the Burley Belt; nearly half of Kentucky's burley tobacco production has been based in the Bluegrass Region, which lies at the heart of central Kentucky.[14] I began my first interview with Martin Henson, whom we will meet in chapter 1, with my standard interview opening. I noted that I was "talking with Martin Henson . . . in Franklin County, Kentucky, about tobacco." Martin corrected me: "Burley tobacco." Although other types of tobacco are grown in Kentucky—particularly dark air- and fire-cured tobaccos—burley has long been both economically and culturally dominant.[15] Central Kentucky farmers are not just tobacco growers. They are burley tobacco growers. This book is therefore specifically about the changed meanings of burley tobacco in central Kentucky as I came to understand and interpret it based on a combination of ethnographic fieldwork and rhetorical analysis.

A Narrative of Change

I am indebted to the first tobacco farmer I interviewed as I began this research in 2005, the late Robert Taylor of Bracken County, in part because he provided me with the knowledge I needed in order to ask my future interviewees much more informed questions than those that I asked him. But he also asked me an important question. Following a driving tour of the Taylor farm that included his cattle and pastures, his tobacco barns, his garden, and of course his tobacco crop, we sat down in the living room for an interview and began with this exchange:

> Author: And one of the things that, that you talked about [during our farm tour] is the different parts of the process? And I wondered if you could walk me through like a whole year, you know, what the different . . .
>
> Robert Taylor: Okay, I'll do my best. Would you rather have present day or olden days? There's a whole lot of difference.

I replied that I wanted him to tell it in the way that he thought it should be told and that perhaps he might tell me about both.

Mr. Taylor's question stayed with me, but it was a while before I realized that he had given me the narrative structure of the story of tobacco in the early twenty-first century: a narrative woven together out of threads of change. While most farmers didn't ask me if I wanted to hear about "present day or olden days" as Bob Taylor did, most offered me a comparative narrative that included how things are done today and how they were done during other periods. Many also shared both questions and predictions about tobacco's future. The tobacco buyout of 2004, which—as I will discuss—ended the federal tobacco price support program that had been in place since the New Deal era, is widely understood as a dramatic moment of change and transition for Kentucky agriculture. Although *transition* in its current usage most often refers explicitly to the aftermath of the buyout, tobacco production can more generally be understood as tradition-in-process, ever changing, continually transitioning. This book unravels multiple threads of change within the lifetimes of present-day tobacco farmers in order to understand how those threads have been woven together into a complex whole.

For Kentucky burley tobacco growers, tobacco farming is a livelihood that involves a mastery of traditional skills passed through generations and adapted to changing circumstances—technological, economic, social, and political. Over the second half of the twentieth century and into the twenty-first, there have been rapid technological changes on the farm, local labor has largely disappeared, women and men have increasingly found jobs off the farm, the federal tobacco program has ended, acreages have grown—substantially in some cases—and the buying habits of the tobacco companies have increasingly moved offshore. During this same period, public awareness of the dangers of smoking has skyrocketed, smoking rates have plummeted, and smoking in public places has become socially unacceptable and in some places illegal. All of this has led to dramatic changes in the political, economic, social, community, and even personal meanings of tobacco. According to D. Wynne Wright, "For public health advocates and social advocates, tobacco is a menace to public health and welfare—a commodity to be stamped out. For farmers, tobacco is a health hazard but, at the same time, an economic and social opportunity wrought by a rich historical legacy."[16] The confluence of the changing symbolic and economic valuing of tobacco must be understood in order to understand the past and present situation on Kentucky farms and to ensure a healthy farm economy in Kentucky's future.

I did not fully understand this for some time.[17] I initially assumed that the narrative of change suggested by Bob Taylor's question was about nostalgia for the loss of the old ways of tobacco production. This in turn reinforced my focus on the cultural and symbolic valuing of tobacco. Farmers set me straight. For instance, Roger Quarles explained in an interview that tobacco was only one part of his farm operation. I asked, "But did it—was it important to you to continue growing some amount of tobacco, despite other opportunities that you had?" He responded, "Well . . . You gotta understand I never did particularly *love* growing tobacco, I did it because it was a business opportunity."[18] I realized that perhaps I wanted to be told about an emotional attachment to the crop, not *just* an economic one. While many other farmers told me that they do enjoy raising tobacco for a number of reasons, I was told many times that no matter how much anybody might enjoy it, no one enjoyed it enough (or was stupid enough) to keep doing it if it didn't pay. Farmers taught me over and over that they continue this tradition because it provides an income.

Farmers emphasized the economic role of tobacco because as the crop has become stigmatized, widespread awareness of the present-day economic importance of tobacco to families, communities, and the Commonwealth of Kentucky is largely gone. I was in the offices of the Kentucky Department of Agriculture (KDA) one day in the summer of 2007 for a meeting in preparation for the Kentucky Folklife Festival, and in a hallway I noticed a display of large, colorful photographs of Kentucky agricultural products. As I scanned the images, I realized there were no photographs of tobacco. Later I visited the KDA website; there were no images of tobacco there either. How had tobacco, once so important to the Kentucky agricultural economy, lauded as central to Kentucky heritage, come to disappear? I wondered. And when had it disappeared? These questions became a central thread that I sought to unravel.

Today tobacco is widely understood as part of Kentucky's heritage—the state's past, not its present. When farmers and I discussed the importance of documenting this tradition for the historical record through oral history interviews, they wanted me to know that tobacco continues to be of economic importance in the present. This connects back to the narrative of change and my assumptions about nostalgia. Farmers cannot simply lament changes within the tradition of tobacco farming, because such changes have helped them to raise the crop more efficiently, and they are therefore necessary to their very survival. However, by asserting that tobacco is of great economic importance, I do not argue that tobacco is not also of great symbolic importance. Rather, the two are intricately entwined. There is nostalgia around tobacco production, but it does not necessarily follow the single path that I initially assumed, for, as Ray Cashman argues, "not all nostalgias are the same."[19] Nostalgia takes multiple forms depending on the position of the person expressing it, as I will discuss.

This book is structured in three parts in order to provide a holistic interpretation of the multiple, changing contexts of tobacco production—changes that are often alluded to but that have not previously been elucidated and interpreted. In part 1 of the book, I trace the "thirteen-month" tobacco production cycle based on my fieldwork with tobacco farmers during the 2007 crop year. The three chapters in this section provide a first-person account of my interactions in the field, combined with quotations and paraphrases from recorded interviews and fieldnotes. I include descriptions of changing cultural practices and circumstances at each stage of tobacco production, as they were described to me. Martin Henson, my most important teacher through the tobacco year, serves as the major guide throughout these chapters, and what I learned from many other participants in my fieldwork is included in order to demonstrate the diversity of perspectives and farming operations. This section is not meant to serve as a comprehensive history of changing farm technologies but rather to provide farmers' perspectives on and experiences with the changes they have seen in their lifetimes. Despite stereotypes of farmers as resistant to change, tobacco farmers have long been accepting of change—when changes prove to be in their economic interest. Rather than merely mourning the loss of "the olden days," farmers understand change and transition as part of the tradition of raising burley tobacco that includes gains and losses.

In part 2, I examine the changing political context of tobacco in Kentucky through an analysis of the Kentucky Department of Agriculture newsletter in order to understand the rhetorical decisions made by the state in response to the changing political status of the tobacco industry and therefore of tobacco production. What began as a search for evidence of the inclusion of images of tobacco production in previous periods in order to find out if the absence I noted in the KDA offices and on the website was indeed a documentable change led to days and days in the Kentucky Department for Libraries and Archives reading every issue of the KDA newsletter, from when it began in the 1940s through the period of my research. I chose to focus on the newsletter because it represents a purposeful public articulation of the priorities and perspectives of the state agency most responsible for shaping the image of Kentucky agriculture. As I took careful notes about what was reported in the newsletters and how this news was framed, the pattern of the shifting politics of tobacco in different periods emerged. It is for this reason that, while I set out to write an ethnography, I have ended up with a book that devotes nearly a third of its pages to the rhetorical analysis of a single printed source.

As the state's official agriculture agency, the KDA generates and promotes the agricultural face of the Commonwealth, and the newsletter has served as the agency's primary means of intentional communication with the public. Along with news, the publication provides the agency's—and therefore the state's—shifting arguments about what Kentucky agriculture is and should be. Over the second half of the twentieth century and into the twenty-first, the KDA's representation of tobacco evolved from an economically vital crop to a celebrated symbol of the state's heritage to a crop replaced by diversified agriculture. By the turn of the new century, tobacco had become a stigmatized crop, and it was no longer politically expedient to claim it as important to the economy. Tobacco is now viewed as heritage, a terministic screen that suggests its economic value is in the past.

Taken together, parts 1 and 2 of this book demonstrate diverging realities: although tobacco production continues in Kentucky, tobacco has become a stigmatized crop, and its existence in the present has largely been erased from public awareness. The occupational and identity category *tobacco farmer* has become, for many, a stigmatized category—what Erving Goffman described as a "spoiled identity."[20] In part 3, I examine aspects of what this means to tobacco farmers. The combined impact of the technological changes and the changing status of the crop and those who grow it has affected the traditional "pride" farmers take in their crop. This is evident in particular expressions of tobacco nostalgia that communicate feelings of loss for a better time of tobacco production, next to the idea that, for some, "*now* is the good old days" because of technological innovations and improved efficiency. Not only was there more pride in tobacco in farmers' fathers' and grandfathers' days, but there was also more respect for the occupation. Tobacco farmers long not for a return to earlier times and technologies—which would be economically unfeasible—but for the pride and respect once associated with a "tobacco-man" identity.

In the final chapter, I examine competing perspectives on the "transition" in which Kentucky burley farmers find themselves. The dominant perspective on the future of Kentucky agriculture is that tobacco production is in its last days and that the "transition" currently taking place is one in which tobacco farmers have replaced, will replace, or should be replacing tobacco production in favor of "diversified" agriculture. This rhetoric suggests that simply planting another crop or raising alternative livestock will lead to the replacement—economically as well as symbolically—of tobacco. This chapter complicates these assumptions, as well as assumptions about the concept of "tradition," through a discussion of tangible and intangible challenges to "replacing" tobacco.

In order to establish a historical context for tobacco production today, it is necessary to provide a brief metahistory of the development of the tobacco industry in the United States and Kentucky as it has been told by past chroniclers. A number of volumes address the history of tobacco in the United States, at least through the mid-twentieth century. In providing an overview of this history from commonly cited works, I am as interested in how the historical narrative has been told as I am in the history itself. Hayden White has argued that the historian "emplots" a particular story of history by pulling from a "chronicle of events" in the historical record.[21] These tobacco histories are all similarly emplotted; they hinge on the same historical events and analogous descriptive passages and end up telling similar uncritical stories about the role of the crop, those who grow it, and those who manufacture products with it.[22] Following this metahistory, I describe recent events that affect tobacco farmers (essentially emplotting my own narrative), informed by media accounts, interviews, Extension Service materials, and recent publications.

Burley Tobacco Production: A Metahistory

There are particular events and details that historians of tobacco repeatedly use to demonstrate the importance of tobacco in the establishment of the American colonies. Tobacco has been called "America's oldest industry" because of its economic role in America before and since the European discovery of the plant upon first contact with Native Americans, at which time it had long been a major item of trade between Native American peoples.[23] As early as 3000 B.C.E., Native Americans "were smoking tobacco for a variety of ritual, social, and diplomatic purposes as well as for personal pleasure"[24] and were also using the plant for a number of medicinal purposes throughout the Americas.[25] Christopher Columbus first mentioned "dry leaves" that appeared to be of great importance to the Indians in a diary entry written in 1492,[26] although the first gift of tobacco he received was said to have been thrown overboard since he and his crew did not know what it was or what to do with it.[27]

Over the course of the sixteenth century tobacco spread across Europe, Asia, and Africa "largely through the agency of traders and sailors who carried the weed and the habit of using it in various ways throughout the world," and by 1607 the Spanish had a "virtual monopoly" on the crop.[28] By this time, the English had developed a "ravenous appetite" for tobacco, despite the admonitions of King James I, making the cultivation of tobacco under English control particularly desirable in order to avoid importation costs.[29] The survival of Jamestown, after two failed attempts, is attributed to John Rolfe's successful development of the crop,[30] opening the way for further "settlement" of the continent. Rolfe made his first attempt to grow a crop of tobacco in 1612, and in the year 1618 Virginians raised twenty thousand pounds.[31] By 1664, twenty-four million pounds of tobacco were being exported from the colonies to England.[32]

Tobacco warehouses were "one of the first businesses to be regulated" in colonial North America.[33] Not only did tobacco help to fund the American Revolution, but some "argue that the unfavorable terms of trade and heavy debt burden that colonial tobacco planters had with English merchants and tobacco consignees were important factors in establishing colonial rebelliousness toward Britain."[34] George Washington raised tobacco,[35] and in 1791, during his first presidential term, tobacco exports totaled $4,359,567, "making it the nation's principal export crop."[36] "The tobacco leaf was woven so deeply into the fabric of American life," writes historian Susan Wagner, "that it was used as a motif in the decoration of columns in the Capitol."[37]

In the earliest days of the Commonwealth of Kentucky, receipts given as proof of stored tobacco "could serve as currency in payment of fees, fines, forfeitures, and debts both public and private."[38] According to John van Willigen and Susan C. Eastwood, "Tobacco is American. Some farmers see it as a link they have with Native Americans. In some regions of America, tobacco is a historic icon. The seal of the city of Lexington, Kentucky, has a tobacco leaf on it,"[39] as does the seal of the city of Owensboro and others.

The Development of Kentucky's Number-One Cash Crop

Europeans learned the basic method of raising a crop of tobacco from Native Americans, "including the details of proper spacing in the field, topping and suckering the plants, and the distinctive drying processes now known as air-curing, sun-curing, and fire-curing."[40] Detailed descriptions of the practices involved in raising a crop of tobacco have, since the eighteenth century, been central to telling the story of tobacco. In 1784, British traveler John Ferdinand Smyth published a two-volume account of his adventures in America that includes a lengthy description of the process of raising a tobacco crop in Virginia.[41] This description is in many ways consistent with methods of tobacco production either as they exist today or at least as they existed within the lifetimes of current tobacco farmers. This includes vernacular language that remains in use today and that will be defined in part 1, such as preparing *plant beds, topping* and *suckering,* the use of *tobacco sticks* in the *curing* of tobacco, *tying* cured tobacco into *hands,* and storing it in *bulks.* Curing practices vary much more widely from then to now, as multiple classes and types have since been developed, each with its own unique curing method and structure. However, the basic idea of curing tobacco using heat and/or air has long been an important step in tobacco production.

A much longer account was provided by William Tatham in his *Essay on the Culture and Commerce of Tobacco,* published in 1800 based on his two decades of observation, beginning in 1769 at the age of seventeen. His account is relevant not only as the most extensive account from the period but because the production and marketing methods that he observed were those that were practiced as tobacco farmers were beginning to settle in what, in 1792, became the Commonwealth of Kentucky. As planters moved west to what is now Kentucky, "the broad outline of the cultural technology of Burley leaf had already become established in Virginia and was carried whole into Kentucky."[42] In addition to farm practices, the complex system of state control of the sale of the crop was "carried whole" into the territory.

The system of slavery was also carried into the territory. As early as 1751, as the territory was being explored, "blacks and whites entered Kentucky together,"[43] and Daniel Boone brought slaves along on his explorations of the territory in the 1770s.[44] The labor of slaves on colonial plantations is central to the story of tobacco in American history. Beginning in the late seventeenth century, a number of factors led to a shift from white immigrant servants to African slaves in the major tobacco-producing region surrounding the Chesapeake Bay. By 1700, farmers in the Chesapeake region were wholly dependent on the labor of slaves in the production of tobacco.[45] According to Joseph C. Robert, a mid-twentieth-century tobacco historian, tobacco "created the plantation pattern. Its labor requirements soon meant hordes of African slaves. Present-day rural and racial problems below the Mason and Dixon Line are rooted in that first Southern staple, tobacco."[46]

With this in mind, the dependence on slave labor seems oddly missing from Kentucky history as it has been written. This is in part due to the fact that "after the Revolution, when nearly all the good land in piedmont had been taken up . . . dissatisfied poor farmers had to leave for Kentucky to find greater opportunities."[47] Many of the farmers who moved to Kentucky and came to depend on tobacco income were small, poor farmers who did not own slaves; Kentucky's was not on the whole a plantation culture. However, as Ann E. Kingsolver notes, the Kentucky historical narrative often ignores Kentucky's slaveholding past, focusing instead on Kentucky's status as a border state during the Civil War, and tobacco "has been glossed as a family-based cash crop."[48] According to Steven A. Channing in *Kentucky: A Bicentennial History,* although "it would be misleading to overlook the feature that most distinguished Kentucky from the lower South, namely the absence of a substantial number of very large plantations," "it is possible to exaggerate the importance of that comparative difference. Apologists developed a powerful mystique around it, using that to argue that slavery was relatively inconsequential."[49] While slavery was a much larger part of the history of the western portion of the state—where the land is flatter and farms larger, and the culture is often described by Kentuckians as more "southern"—there were slaves on farms across the state, including the Bluegrass Region, which became the center of burley production.

The importance of burley tobacco, which had become dominant in Kentucky as it spread to the central region, grew as chewing tobacco became the most widespread method of consumption in the United States because burley was a primary ingredient. At this time, new strains of red burley were being developed, and leaf quality was improving. The move to "chaw" during the first half of the nineteenth century was fostered in large part by the desire on the part of Americans to separate themselves from what were seen as elite and effeminate European ways. As the masculine hero became the frontiersman, a "man of manly independence,"[50] and the "common man" "reigned supreme," chewing spread up the class ladder.[51] Chewing tobacco was the "only one of our tobacco customs which did not originate in the conscious imitation of European manners."[52] Although tobacco was grown in Kentucky in the early nineteenth century, it was not until the 1830s, when a canal was built in order to provide a consistent and safe route around the Falls of the Ohio at Louisville, that Kentucky began to take its place as a major producer of tobacco as well as hemp.[53] Even then, however, tobacco production continued to center around the rivers, and very little was produced in the Bluegrass Region.

By the mid- to late nineteenth century, distinct types of tobacco with different production practices and uses had settled into particular regions based on climate and soil conditions. Cigar tobaccos (binders, wrappers, and fillers) remained concentrated in the North (the Connecticut Valley, small portions of New York and Pennsylvania), as well as small parts of Georgia and Florida. Flue-cured or bright tobacco became concentrated in the Carolinas and Georgia. Dark fire-cured and dark air-cured tobaccos—used primarily in chewing tobacco and snuff, as well as pipe tobacco—settled in far western Kentucky and northwestern Tennessee. According to John Morgan, the practices that are still used in the production of dark fire-cured tobacco remain the most consistent with tobacco production of colonial times.[54] Burley settled in Kentucky and parts of Tennessee, and Maryland retained its own air-cured type. Today, the overwhelming majority of burley, Maryland, and flue-cured tobaccos is used in cigarettes. A small region of Louisiana, centered in St. James Parish, became the home of a particularly specialized type of pipe tobacco, Perique tobacco.[55]

According to numerous versions of the tobacco history narrative, "warfare has been the single most significant influence on the worldwide propagation of a taste" for tobacco.[56] The Civil War had both direct and indirect effects on Kentucky's move to the top of tobacco production. In stories told about Union and Confederate soldiers meeting in the darkness between battles to exchange news and provisions, it is said that Confederate soldiers traded their chewing tobacco for Union coffee because of the shortages of each in their respective regions.[57] As a result, the northern appetite for chewing tobacco grew.

The Civil War also shifted the important tobacco regions. Because Kentucky was a border state, it was largely spared the structural devastation of states that had joined the Confederacy. This meant that during and after the war, Kentucky tobacco farmers had a distinct advantage over their counterparts in states such as Virginia and North Carolina, where farms and warehouses had largely been destroyed, as had centers of marketing and manufacturing.[58] Louisville, the center of the market of the West at the time, also came through the war relatively unscathed, unlike potential competitors such as Nashville, Atlanta, and Birmingham.[59] Meanwhile, New

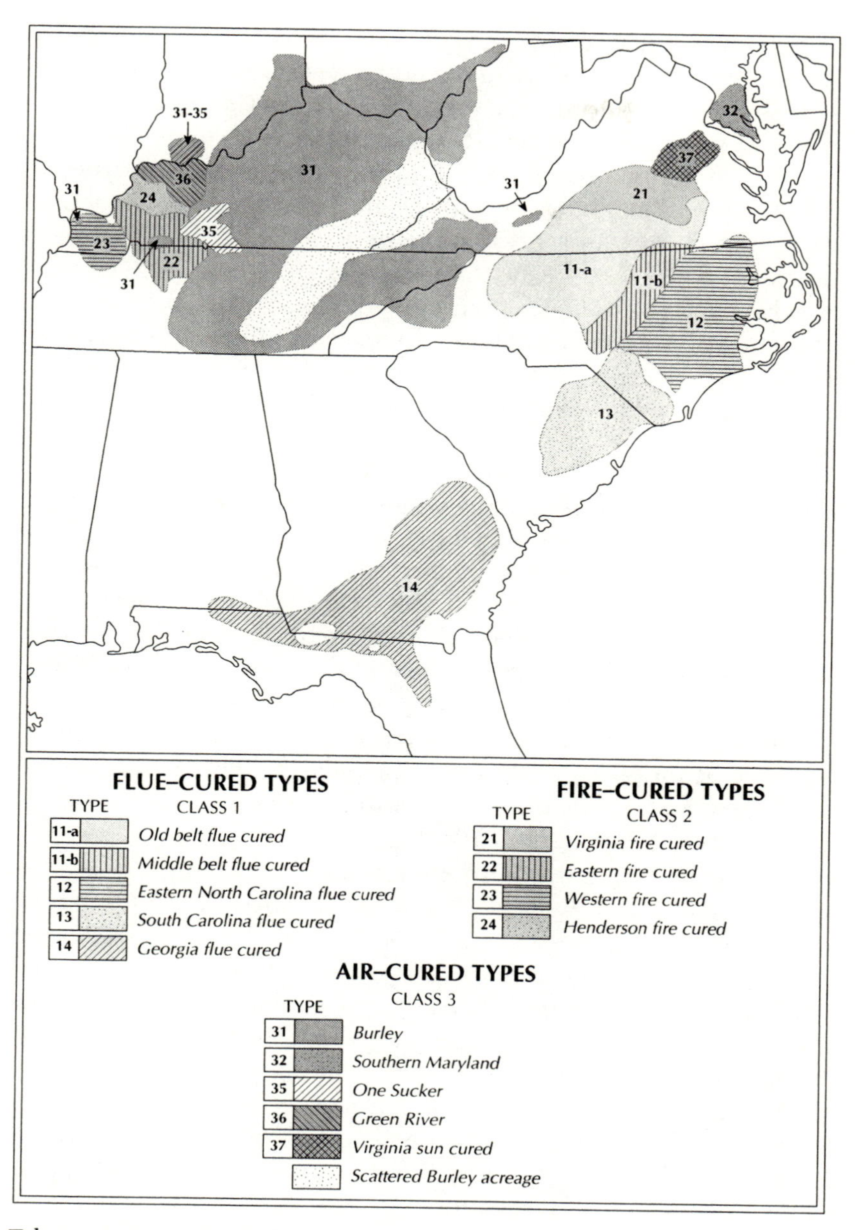

Tobacco growing regions, historically, by class and type (USDA). (Map drawn by Richard A. Gilbreath)

York became a marketing center for the sale of tobacco produced in Union states.[60] Although Virginia never recovered its former share of tobacco marketing and manufacturing, North Carolina eventually did recover and became the nation's largest tobacco-producing state.

Another development just before the war ended was instrumental in establishing Kentucky's place in the tobacco industry ever more firmly—the discovery of white burley. The burley tobacco leaf that was being grown at the time was a harsh red variety. The story of white burley—Kentucky burley's origin narrative—is recounted in nearly all published histories of tobacco, with an accompanying explanation that it was one of the rarest of moments in the natural world, called a "sport" or "a sudden deviation from a standard type."[61]

The story goes that the tenants of a Brown County, Ohio, tobacco producer named Fred Kantz ran out of seed and bought some from a Bracken County, Kentucky, farmer named George Barkley. The tenants planted the seeds in their seedbeds, and when it came time to transplant the plants into the field, they didn't look right—they were "dirty yellow in color," although "sturdy enough"—so they were destroyed.[62] The next year, one of the tenants, George Webb, planted leftover seeds that grew to become leaf that "cured out to a handsome, almost golden, light tan or cream leaf."[63] The following year, he raised twenty thousand pounds; the tobacco was praised by buyers, and it took off and "positively thrived" in the Bluegrass Region[64] beginning in the 1870s.[65]

Between 1865 and 1929, Kentucky produced more tobacco, by pound, than any other state.[66] White burley required a shorter growing season, the entire stalk could be harvested at once (*stalk-cut*) rather than requiring multiple passes through the field harvesting the leaves as they matured (*primed*), and it could be air cured rather than fire cured. It was appealing to manufacturers because it had characteristics that made it ideal for chewing and pipe tobaccos: there was less sugar in the leaf, so it absorbed sweeteners that were added to it for flavoring and that helped it to blend well with other tobaccos. These characteristics would later make it an essential part of the American-blend cigarette.

Changes in tobacco marketing took place from the 1830s through the 1890s, as auctions became "institutionalized," and farmers began to move away from packing their tobacco in hogsheads.[67] By the beginning of the nineteenth century, there was growing distrust of the "tobacco note" that had long served as the representation of a hogshead of tobacco as it was bought and sold, and by the 1830s the auction system was developing as buyers began purchasing directly from farmers immediately following the inspection of their tobacco.[68] Looseleaf sales were introduced in Kentucky and Tennessee in 1901, replacing sale in hogsheads.[69] The practice of packing hands of tobacco onto baskets at the warehouse, rather than in piles on the floor, was established in Lexington in 1904[70] and would remain the standard marketing practice until the early 1980s; even today, this practice remains the iconic image of the tobacco sale. The last hogsheads sales in the burley region took place during the 1929–1930 marketing season in Louisville.[71]

The Tobacco Trust, Trouble in the Black Patch, and the Tobacco Program

Tobacco overproduction was identified as a problem immediately following John Rolfe's first planting of the crop in Jamestown in 1612, and with the concept of overproduction came governmental monitoring and intervention. As early as 1613—just the second year of cultivation at Jamestown—Deputy Governor Thomas Dale feared yet another failed colony and "ordered that no man could raise tobacco unless he also each season manured and maintained two acres of corn."[72] According to William Tatham, as early as 1620 King James I ordered colonists to limit tobacco production and instead to plant corn and potatoes and raise livestock,[73] and the first legislation ever passed regarding tobacco commerce, in 1639, addressed the need to limit tobacco production.[74] Overproduction was at the heart of the 1670s event known as Bacon's Rebellion, and in 1682 there were "plant-cutting riots" in which planters cut their own and their neighbors' tobacco when the Virginia Assembly refused to impose production limits.[75]

While the relationship between farmers and the tobacco industry had been a tumultuous one since the strife between planters and British merchants, the conflict intensified beginning in the 1870s as manufacturing became consolidated. Until the late nineteenth century, tobacco products were largely produced by "country factories"—everyone from merchants to planters produced chew for sale locally.[76] Gradually, cities such as St. Louis and New York, as well as towns in North Carolina, became centers of manufacturing.

The American Tobacco Company was established in 1890 "under the guiding hand of J. B. Duke,"[77] the son of a poor planter made rich through home manufacture and traveling sales of tobacco products beginning just after the Civil War.[78] "The Trust," as it came to be called, was made up of the largest tobacco manufacturing companies, and it rapidly achieved a monopoly in the industry through rampant price wars that drove their competitors "not out of business, but into joining the Trust," which ultimately subsumed over 250 manufacturers.[79]

The formation of the Trust, along with other factors, resulted in severe price reductions, and farmers attempted to overcome shrinking farm incomes through increased production. Increased production led to overproduction, which served to drop the prices further. One result was the "Black Patch Wars," which most famously took place in the dark tobacco region of far western Kentucky and Tennessee.[80] Dark-tobacco growers experienced the most drastic price reductions as the demand for their product dropped when domestic tobacco consumption began to move away from chewing tobaccos to smoking tobaccos at the end of the nineteenth century. The first attempt to build a cooperative association to fight the growing power of the Trust took place in this region in 1904, with what became the Dark-Fired Tobacco District Planters' Protective Association of Kentucky and Tennessee. Alongside this organization, a secret association was formed that became known as the Night Riders. Beginning in 1906, the Night Riders organized themselves "with robes and masks, and [with an] elaborate paramilitary hierarchy operating as an outlaw underground army" that "coerced reluctant leaf planters to join . . . flogged still others, dragged plant beds, burned barns and houses, killed some."[81] They also burned warehouses and manufacturing facilities in the towns of Princeton and Hopkinsville, Kentucky.

Although the activities of the Night Riders in the dark-fired region have been most widely documented, there were similar movements in central and northern Kentucky as well. The tactics of the Night Riders in other regions were reportedly "more successful and somewhat less violent,"[82] but oral accounts of violence and murder have been passed through families. The activities of the Night Riders ended around 1908 in large part because of legal action taken against them by their victims, but also because by this time public sentiment had turned against them. In 1911, the Trust was dissolved by the US Supreme Court as a violation of the Sherman Anti-Trust Act of 1890. However, out of the Trust came the "Big Four" companies—American Tobacco Company, Liggett and Myers, Lorillard, and R. J. Reynolds—and the turbulent times were far from over.

Early twentieth-century cigarettes—a form of tobacco use that was not yet widespread—were primarily "Turkish-blend" cigarettes. This blend comprised about 60 percent domestic tobaccos—but not burley—and 40 percent Turkish tobacco, a type not raised in the United States. R. J. Reynolds introduced its Camel brand of cigarettes in 1913, and with it and the competing brands that followed came a cigarette blend that "revolutionize[d] the cigarette field."[83] Though particular blends have been trade secrets over the years, this "American-blend" cigarette generally included—and still includes—about 50 to 60 percent flue-cured or bright tobacco, 30 to 35 percent burley, 10 percent Turkish, and about 2 percent Maryland leaf.[84] Camels were followed by the American Tobacco Company's Lucky Strikes in 1918 and by Liggett and Myers's Chesterfields in 1919, all American blends. With these brands, the companies ushered in the era of "concentrated one-brand advertising."[85] This new blend also increased the demand for burley tobacco.

The First World War helped to spread these new cigarettes, and tobacco prices soared to all-time highs, ranging from twenty-five to thirty cents a pound.[86] This height, however, was followed by a bad crop year in 1920, and prices plummeted to an average of thirteen cents a pound in the Lexington burley market. Once again, farmers were angered and determined to band together in hopes of pressuring the companies for higher poundage prices, and during this period cooperative associations were formed for each of the tobacco classes. The Burley Tobacco Growers Co-operative Association was formed in 1921 with a membership goal of 75 percent of all burley growers. These efforts were successful for a few years, with prices reaching over twenty-eight cents a pound in 1922,[87] but by 1926 the efforts were failing, the crop was not cooperatively managed, and the price dropped back to twelve and a half cents.[88] As the Great Depression hit, not only were prices low, but demand for cigarettes fell, and by 1931 burley brought about eight and a half cents a pound and dark-fired three cents.[89] Attempts to revive cooperative associations during this period failed.[90]

With "Franklin D. Roosevelt's eventful first hundred days of New Deal legislation" came the Agricultural Adjustment Act of 1933 (known as the AAA), which "focused on wheat, cotton, field corn, hogs, rice, milk, and tobacco and provided for restricted production and benefit payments to the farmer."[91] This act was struck down in 1937, and a new version was passed in 1938;[92] various amendments were added over the years. With the passage of the Agriculture Adjustment Act of 1938, all tobacco growers were "permitted to vote through referendum [every three years] to impose production quotas in return for a supported price."[93] The US Department of Agriculture (USDA) was also charged with creating a system of inspection and outlining a structure for the uniform grading of tobacco by government graders through the Tobacco Inspection Act of 1935,[94] which in essence strengthened 1916 and 1929 legislation by making uniform grading and inspection mandatory at no cost to the farmer.[95] The tobacco program generally stabilized tobacco production for decades, even through times such as the Second World War, when war once again led to a boom time for tobacco. American cigarette consumption rose 75 percent between 1939 and 1945,[96] and the Second World War spread the American cigarette around the globe.

Those who had grown tobacco in the years leading up to the passage of the AAA were given a "base" or "allotment": a precise number of acres or part of an acre of tobacco that could be grown and sold on each farm without penalty, based on how much had been raised on that farm in the years prior to the program. This base was then adjusted annually dependent on the projected demand of the tobacco companies (both domestic and exports) and the amount of tobacco in the pool stocks held by the cooperatives, and support prices were set for each grade. Because tobacco is aged approximately three years before it is used, company estimates were based on their projected needs three or more years in the future. When it came time for farmers to sell their tobacco at auctions that took place at tobacco warehouses, if buyers—either representing specific tobacco companies or "leaf buyers" who bought tobacco for multiple manufacturers worldwide—were not willing to bid at least a penny above the support price, then that tobacco went to a "pool" managed by a grower cooperative. The Burley Tobacco Growers Co-operative Association was revived in 1941 to manage the burley pool stocks, paying farmers for tobacco that was not bought at auction with money borrowed from the Commodity Credit Corporation of the USDA and later paid back with interest when the tobacco was sold. Farmers and others on the production end are quick to point out that unlike other commodity programs, the tobacco program was not a subsidy program, and although there were administrative costs associated with it, the government more than gained these costs back through interest gained on loans to the pool, a topic to which I will return.

It has been argued that tobacco production techniques and marketing procedures, unlike those of commodities such as cotton, changed little as a result of the AAA.[97] However, acreage allotments meant that farmers could sell every leaf of tobacco that they raised on their allotment, and they therefore resulted in ever-growing average yields through not only the careful collection of each leaf but also the development of new production techniques and varieties and the application of new synthetic fertilizers. Average yields tripled between 1939 and 1971. The research that led to the sharp increases in yields was described by one longtime Burley Co-op president, John Berry Sr., as "cruel economics and blind scientific endeavor"

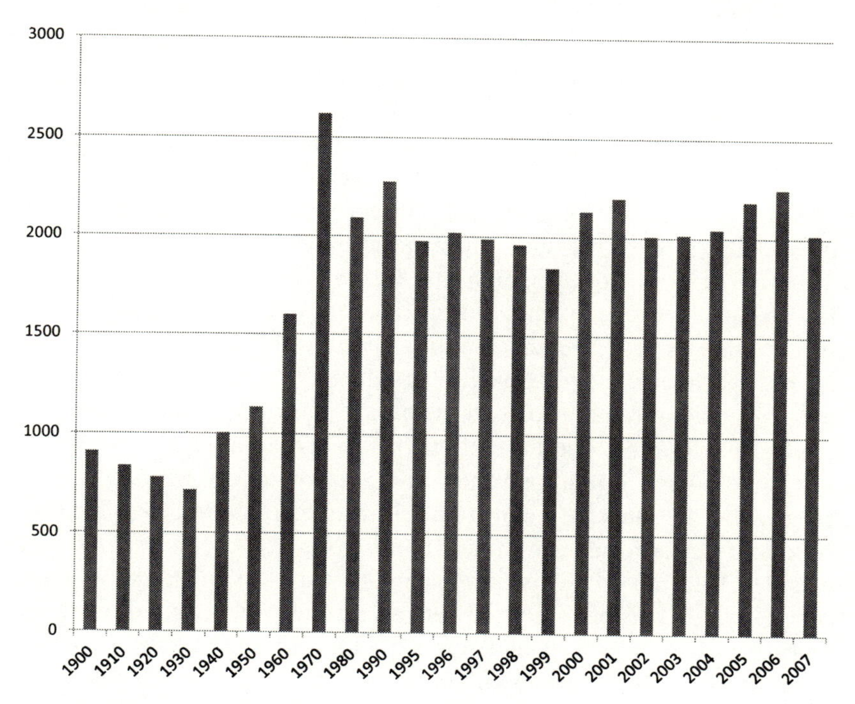

Average yield per acre (in pounds), Kentucky tobacco, all types. Data provided by decade through 1990 and annually beginning in 1995. (USDA, Economic Research Service)

because higher yields led to more tobacco at lower prices, ultimately leading to more work for the farmer at lower wages.[98] The industry continued to be plagued by overproduction, and quotas rose and fell over the years.

In 1971, when there was once again too much burley being raised and an abundance of tobacco in the pool stocks, growers passed a referendum that changed the acreage quota system to a poundage system; at the same time they agreed to a quota reduction.[99] This meant that farmers could plant an unlimited number of acres in order to raise their poundage quota of cured leaf. This is understood by many as one link in a chain of events that decreased the care with which each individual leaf was treated and therefore changed traditional practices designed to preserve the full value of each leaf.[100] Yet during the push for a move from acreage to poundage the Kentucky Department of Agriculture argued that this shift would result in *better*-quality tobacco because farmers were overusing fertilizers in order to get the most out of their allotted acreage.[101] However the shift is interpreted, yields gradually dropped to an average of 2,100 pounds,[102] although 3,500 pounds to the acre is not unheard of today.

For most of the years of the program, a tobacco marketing quota could almost exclusively be obtained through the purchase of land that came with a quota, which meant that land values were heavily influenced by the amount of tobacco base that came with a piece of land. There had long been a sharecropping and tenant system in which growers engaged in various arrangements, such as *on shares* (a relationship in which one farmer raised another's tobacco for a share of the crop that varied depending on who supplied the inputs such as seed, fertilizers, equipment and barns, labor, etc.) or *on halves* (a sharing relationship in which the landowner and tenant split the inputs and profits evenly). Beginning in 1971, farmers could lease quota owned by others; these arrangements also varied, but by 2004 some paid as much as eighty or ninety cents a pound to grow someone else's allotment. Lease costs varied significantly by county, according to demand, and were highest in central and western Kentucky.[103] This also meant that a class of nonproducing quota owners was born, so that retired farmers, widows, nonfarmers who bought a farm that had some quota, and even businesses and institutions such as schools and churches came to depend on leasing their tobacco allotments as a source of income. Beginning in 1991, it was possible to buy quota separate from a parcel of land within the county in which you lived.[104] The ability to lease or buy quota in other counties, known as *cross-county leasing*, was a subject of contentious debate right up to the end of the program in 2004; it was voted on numerous times by growers and never passed in Kentucky. The primary argument against cross-county leasing was that it would benefit only the largest farmers.

Tobacco Threatened

Tobacco use was criticized beginning with first European contact. King James I published *A Counterblaste to Tobacco* in 1604, and when that did not stop the spread of its use throughout England, he imposed heavy taxes. Movements against tobacco use rose and fell in the centuries following King James, and the first "significant" anti-tobacco tract in the United States was published in 1798.[105] Well-known Americans such as Horace Mann, Henry Ward Beecher, Horace Greeley, Thomas Edison, and Henry Ford were anti-tobacco proponents, as were the official organizations of Methodists and Quakers.[106] In 1902, the Quakers condemned "the grant of public money for use in research concerning growing and curing tobacco," and despite what had been an ongoing split between northern and southern Methodists on the issue, in 1914 the Methodist Conference forbade "candidates for the ministry from smoking."[107] Joseph C. Robert quotes a southern Methodist's comments in the Reconstruction period: "Those Northern Methodists . . . are engaged in a general crusade against tobacco. That is our Southern staple, and our churches are largely supported by it."[108] The movement against tobacco use in the United States gained momentum in the early 1900s alongside the Prohibition movement.[109] Several states banned the sale and public use of tobacco, and the anti-tobacco movement even served as the platform of the 1920 presidential campaign of Lucy Page Gaston, a former member of the Women's Christian Temperance Union and a leader in the anti-tobacco movement.[110]

Widespread public acceptance of the negative health effects of tobacco finally began to take hold in the 1950s and '60s. In 1954, "the first cigarette cancer scare coincided with the introduction of the modern filtered cigarette,"[111] which was actually an improvement on existing filters. This scare came in the form of a report released by the American Cancer Society to the American Medical Association, which was taken quite seriously by the public,[112] resulting in a dip in sales and in tobacco allotments.[113] But the major blow to tobacco came with the 1964 Surgeon General's Report on Smoking and Health. The report, issued on a Saturday to avoid immediate repercussions on the stock market, made a connection between tobacco use and lung cancer that was taken more seriously by American consumers than any previous expression of belief about the ill effects of tobacco use.[114] It included the statement that "cigarette smoking is a health hazard of sufficient importance in the United States to warrant appropriate remedial action."[115] This was followed, in 1965, by a required warning label on cigarette packages that stated—as a result of concessions to the tobacco companies—simply that "cigarette smoking may be hazardous to your health"[116] and a ban on cigarette advertising on broadcast television in 1971.[117]

Tobacco companies rushed to buy other manufacturing concerns in order to protect both their image and their finances. Philip Morris led the way, buying up subsidiaries that made everything "from chewing gum and razor blades to beer."[118] Pro-tobacco associations had begun to organize in the late 1910s and early 1920s,[119] and the Tobacco Institute—which would become the major lobbying arm of the tobacco industry—was established in 1958.[120] The Council for Burley Tobacco formed in 1971, because "the challenge of mounting defenses against anti-tobacco attacks promised to be a full-time job."[121] Taxes were viewed as a threat to the industry perhaps equal to the growing awareness of health effects, and increased excise taxes were often successfully fought off. Manufacturers successfully mobilized growers to fight such fights for them, through active campaigns to tie the interests of farmers with their own interests and put a sympathetic face on the industry.

Recent Events: The Master Settlement Agreement, the Tobacco Buyout, and Government Regulation of Tobacco

President Bill Clinton was perceived by many to be "the most anti-tobacco president in history."[122] He paid a visit to Kentucky in 1998 to meet with farmers, assuring them that "we don't have to wreck the fabric of life in your community. We don't have to rob honest people of their way of life."[123] In September 2000, the Clinton administration created the President's Commission on Improving Economic Opportunity in Communities Dependent on Tobacco Production while Protecting Public Health. The lengthy name reflects the complexity of the issues with which it grappled. The commission's report, released in May 2001, made a range of recommendations focused on revamping but not eliminating the tobacco program, providing financial compensation and technical assistance to encourage growers to diversify their farm operations, and supporting tobacco cessation programs.[124] This commission represented one aspect of a movement in the 1980s and '90s to bring public health advocates and tobacco communities together on those issues that they could agree upon, such as the economic consequences of a decline of tobacco markets and working to end youth tobacco use.

The $206 billion Master Settlement Agreement (MSA), reached in 1998 between the four largest American tobacco companies and forty-six state attorneys general, both symbolized and fortified the change in public attitudes regarding tobacco, as tobacco manufacturers settled with states that sought to recover medical costs associated with smoking-related illnesses. The MSA was the largest settlement of a civil suit in American history. Florida, Minnesota, Mississippi, and Texas had made prior, separate settlements totaling $40 billion. The companies originally sued were Philip Morris, R. J. Reynolds, Brown & Williamson, and Lorillard; additional companies signed on subsequent to the original settlement. Kentucky's participation in this lawsuit itself posed questions. Scott White, a member of the Kentucky attorney general's staff, said in an interview for the Kentucky Oral History Commission in 2000, "But of course *tobacco* affects so many different parts of Kentucky's economy and is grown in 119 out of our 120 counties. We've got a manufacturing plant in Jefferson County, there's tobacco warehouses strewn across the state, I mean, it truly is . . . you know, just part and parcel of who we are as Kentuckians. And, so we were, we kind of felt like if we were to sue the cigarette companies, it'd be kind of like Kentucky suing itself." Nevertheless, Kentucky participated in the lawsuit.

In addition to the monetary settlement, the industry agreed to create a fund for the promotion of tobacco prevention and cessation, the American Legacy Foundation, and to "modest restrictions on advertising and promotion," including a ban on the use of cartoon characters such as Joe Camel and of billboard advertising and promotional merchandise.[125] In addition, the industry agreed to the closure of lobbying organizations, including the Tobacco Institute, and released what amounted to millions of pages of internal industry documents. Public health advocates had hoped for many additional outcomes, including legislative changes such as Food and Drug Administration (FDA) regulation of tobacco, stronger warnings on packaging, and tougher restrictions on the promotion of tobacco products. According to Allan Brandt, in his critique of the MSA, it "proved to be one of the industry's most surprising victories in its long history of combat with the public health forces."[126]

The Master Settlement Agreement resulted in two phases of monetary awards. Phase I monies are being paid out to states over twenty-five years, and Phase II monies provided annual payments to growers in order to compensate them for the expected decline in demand for tobacco. The payments to farmers ended with the tobacco buyout, discussed below. Kentucky allocated 50 percent of the state's Phase I monies to the support of agricultural diversification through the newly created Kentucky Agricultural Development Board, discussed in chapter 7. The use of Phase I monies by states has generated widespread criticism in the ensuing years, primarily because so little of the money went to antismoking programs, as public health advocates had hoped. Many states simply used the funds to plug holes in their budgets. "The costs of the settlement, as predicted, were passed on to consumers," as the major companies raised their prices in order to cover the costs.[127] Some argue that states are even more dependent on the tobacco industry for revenue than they were prior to the settlement.[128]

Meanwhile, farmers were experiencing a tobacco program that had become a rollercoaster ride. Throughout the late 1990s, quotas rose and fell dramatically. In 1991, quotas were raised over 20 percent, followed by cuts at or below 10 percent until 1995, when they began to climb again. In 1999, farmers saw a 29 percent cut, followed by a 45 percent cut in 2000. Average quota lease prices increased from under thirty cents in 1997 to nearly seventy cents in 2002,[129] as quota owners attempted to maintain their incomes. In the year 2000, Philip Morris established what it called the Tobacco Farmer Partnering Program, through which it began encouraging growers to bypass the auction system and contract directly with them, and in the following years other companies followed their lead. The controversy and uncertainty that came with contracting, combined with the dramatic quota cuts, served to increase feelings of inevitability surrounding the demise of the tobacco program. The end of the program came in October 2004.

For over sixty years, the pool stocks maintained by the Burley

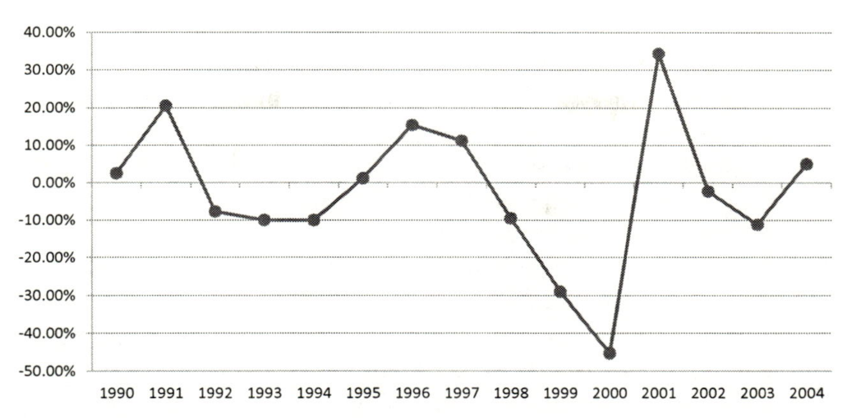

Basic quota, burley tobacco, as it rose and fell, 1990–2004. (William Snell, University of Kentucky College of Agriculture, USDA, Economic Research Service)

Co-op ensured that growers had a market for their tobacco, and the guaranteed minimum price ensured, with a few exceptions, that the price did not drop significantly from one year to the next. However, support prices also ensured that American tobacco was increasingly more expensive than tobacco grown in other parts of the world, such as parts of Africa, South America, and Asia. Between 1970 and 2002, the portion of burley tobacco imported into the United States for domestic use grew from 0.6 percent to 48.1 percent.[130] While American burley growers were being paid about two dollars a pound prior to the buyout, producers elsewhere were raising burley for fifty to seventy-five cents a pound, "maybe up to $1.00/lb in some years depending on the country as well as the year."[131] The program also never fully controlled the problem of overproduction, particularly in bad crop years when unwanted leaf went into the pool in large quantities. This meant that when pool stocks grew excessively large, deals were struck in which tobacco manufacturers bought out the pool stocks in exchange for lowered prices and quotas for farmers. The largest of such buyouts took place in 1985, when farmers accepted a thirty-cent cut in poundage prices in exchange for an agreement in which the companies bought the pool stocks over a period of several years.

During the 1980s and 1990s there were multiple attempts to dismantle the tobacco program, such as an amendment to the 1981 farm bill that came one vote shy of passage. In 1982, "no-net-cost" legislation instituted a system of fees, shared between producers and buyers, to cover the administrative costs of the program in order to ensure that taxpayers were not supporting tobacco production.[132] By the late 1990s, the congressional delegations of Kentucky and other tobacco-producing states were proposing their own legislative solutions to the problems plaguing tobacco farmers and the tobacco industry. By this time, it was clear to all involved that major changes to the system were needed, although there was little agreement on what those changes should be, and Congress was working to please the opposing constituencies of growers, quota owners, manufacturers, and public health advocates. Many proposals were made, some of which became legislation, some of which did not; in 2002 alone, nine tobacco buyout bills were introduced in Congress.[133]

In October 2004, Congress ended the federal tobacco program with the passage of the Fair and Equitable Tobacco Reform Act (within the American Jobs Creation Act of 2004), resulting in "one of the most dramatic changes in any U.S. agricultural policy over the last half century, as tobacco now ha[d] the distinction of being the only government-supported commodity to move abruptly to an entirely free-market policy."[134] The primary argument for the buyout was that once there was no longer a support price, the market would readjust, and manufacturers would buy more American-grown tobacco. It was also argued that the complicated system of nonproducing quota owners and growers to whom they leased their allotments would be simplified, and growers could accept lower prices since they would not have to lease in poundage.

The end of the program is referred to as "the buyout" because quota owners and tobacco growers were entitled to annual payments for ten years, based on the amount of tobacco grown and/or the quota owned under the tobacco program. These payments, totaling $9.6 billion, come from the Tobacco Transition Payment Program, funded by the major cigarette manufacturers, not from tax dollars, as many erroneously believe. The purpose of buyout payments was to compensate growers and quota owners for an anticipated loss of income, as well as the probable decrease in land values once tobacco quotas were no longer attached to farms. Presumably, owners would lose income as they lost the ability to lease their quota, while growers (whether owners or leasers of quota) would lose income because the poundage price would drop once the support price was gone. Those who owned quota at the time of the buyout received payments of $7 per pound of quota owned in 2002. Growers who had leased other people's tobacco quota received $3 per pound grown, and those who raised tobacco in crop-share arrangements rather than leasing another farm's quota straight out received a portion of the grower payment based on that arrangement (i.e., those who raised on halves received $1.50 per pound). Many growers were in multiple categories—owning and raising their own quota while also leasing quota in order to supplement what they owned. As the name—Tobacco Transition Payment Program—indicates, in addition to its compensatory purpose, the buyout was intended to help tobacco producers through a "transition," a term that (as I will discuss in chapter 7) has come to have multiple meanings.

The tobacco buyout is not a true buyout, because tobacco growers did not sell their quota and with it their right to raise tobacco.[135] Although many did get out of tobacco, those who remain now work in a free-market environment with no poundage limits but also no support price. The poundage price dropped from about $2 a pound to around $1.50 the first year after the buyout. For those who were paying seventy to ninety cents a pound to raise tobacco for $2 a pound, $1.50 a pound was clearly a better deal even if they continued to rent land, because without the quota land became considerably cheaper to rent. However, there were also many growers who owned quota—including many who had bought quota after the law changed to allow them to do so in 1991—and these owner-growers lost out (although for ten years their buyout payment serves as at least partial compensation for their losses). The effects of the tobacco buyout are discussed throughout this book.

A number of the buyout proposals of the late 1990s and early 2000s included the granting of regulatory power over tobacco to the FDA—something that health advocates had lobbied heavily for and tobacco companies had long lobbied hard against. Although this power did not make it into the final buyout legislation, it soon followed. FDA regulation of tobacco had long been a topic of great contention. When the FDA was created in 1906, the tobacco industry successfully lobbied for tobacco's removal from the list of drugs to be regulated.[136] In 1996, the FDA declared that it had the power to "regulate nicotine-containing tobacco products as medical devices,"[137] but in 2000 the Supreme Court ruled that the FDA did not have the jurisdiction to do so without congressional action. FDA regulation of tobacco products finally became a reality with the Family Smoking Prevention and Tobacco Control Act, signed by President Barack Obama on June 22, 2009. Passage was made possible in large part by the support of Philip Morris, which reversed its long-standing opposition. Philip Morris, which claims to be conducting research toward "safer" tobacco products, seems to have decided that FDA involvement will shield it from future litigation. It is the perception of many—particularly other tobacco companies—that the limits on tobacco product marketing included in the legislation will help Philip Morris to protect its control of the cigarette market. In 2008 Philip Morris split into Philip Morris International (PMI) and Philip Morris USA (PMUSA), suggesting a connection with its support of FDA regulation, since presumably PMI, which makes products for sale abroad, would not be under FDA regulation, while PMUSA would.[138]

Farmers have long feared—with the encouragement of tobacco companies[139]—that FDA regulation will translate into additional governmental involvement in on-farm practices. Although the legislation bars the FDA from regulating growers, the regulation of the content of tobacco products will certainly lead to more intense regulation—of chemical use and other farming and packaging practices, for instance—of growers by manufacturers. Although growers have long opposed FDA regulation, once it became a reality, some expressed hope that it will result in an increase in the amount of domestic tobacco purchased by manufacturers because "foreign leaf [used in the domestic manufacture of products] will have to meet the same standards for pesticides as domestic leaf."[140] Whatever tobacco farmers' feelings may be, FDA regulation adds another layer of uncertainty about tobacco's future and the ongoing transition of tobacco production.

Kentucky Burley Tobacco Production Today

As I demonstrate throughout this book, tobacco farming as cultural practice has been eulogized in the news and by Kentucky authors, it has been erased from the publications of Kentucky government agencies, it has been removed from local festivals, and it has been put on exhibit as a "way of life" of the past. Informal conversations I have had since I began this research suggest a widespread public perception that there are very few tobacco growers left.

It is indisputable that US tobacco production is in decline, but

Number of All US Farms Growing Tobacco, All Types	
Year	**Farms Growing Tobacco (thousands)**
1954	512
1959	417
1964	331
1969	276
1974	198
1978	189
1982	179
1987	137
1992	124
1997	94
2002	57
2007	16
Source: US Census of Agriculture, Bureau of the Census, National Agricultural Statistics Service, USDA	

Kentucky Farms and Kentucky Farms Raising Tobacco, All Types			
Year	**Total Number of Kentucky Farms**	**Number of Farms Growing Tobacco**	**Percentage of Farms Growing Tobacco**
1920	270,626	143,599	53
1940	252,894	126,691	50
1959	150,986	119,970	80
1978	102,263	73,932	72
1992	90,281	59,373	66
2002	86,541	29,237	34
2007	85,260	8,113	10
Source: US Census of Agriculture, Bureau of the Census, National Agricultural Statistics Service, USDA			

the decline is not a new phenomenon. The number of farms growing tobacco of all types and in all regions of the United States fell from 512,000 in 1954 to 56,977 in 2002.[141] Between 2002 and 2007, the number of farms growing tobacco nationwide dropped from 56,977 to 16,234, although the number of pounds that were raised during the same period dropped only from about 873 million to 778 million.[142] There were nearly 60,000 Kentucky farms on which tobacco was grown in 1992 and nearly 30,000 in 2002.[143] Kentucky tobacco production dropped over 30 percent in 2005, the year following the buyout, but the decline in Kentucky burley and dark-fired tobacco

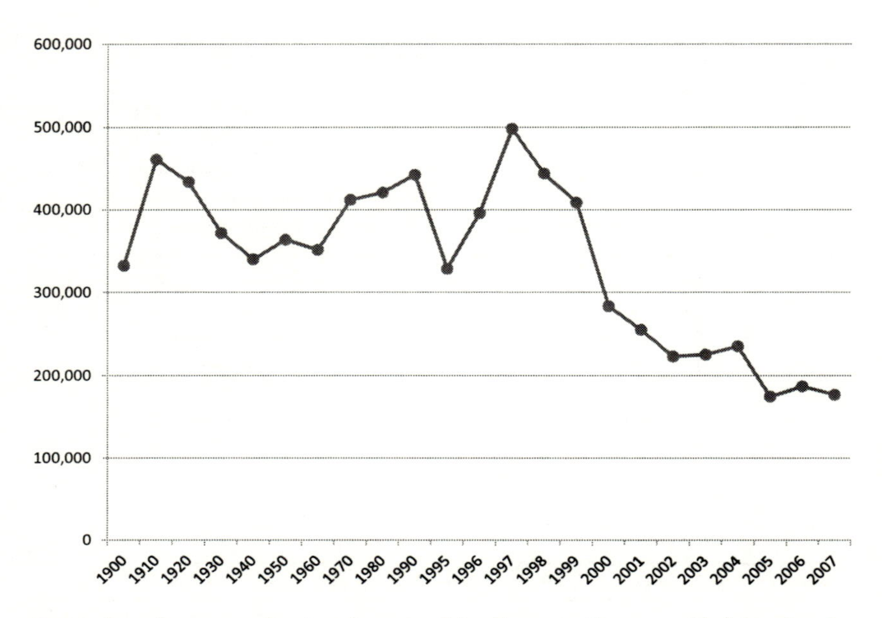

Kentucky tobacco production (in pounds), all types. Data provided by decade through 1990 and annually beginning in 1995. (USDA, Economic Research Service)

production was smaller than the decline in areas that specialize in flue-cured tobacco, such as North and South Carolina and Virginia.[144] Kentucky continues to lead in the production of burley and dark tobaccos, and only North Carolina outranks Kentucky in total tobacco production.[145] However, although Kentucky produced 75 percent of US burley in 2007, this represented just over 15 percent of the world's burley, down from 40 percent in 1990.[146]

A full 50 percent of US tobacco farms are in Kentucky, and in 2007 "the number of farms growing tobacco outnumber[ed] all other single ag enterprises in Kentucky with the exception of the number of cattle/hay farms."[147] The 2007 Census of Agriculture revealed that there were 8,113 tobacco farms in Kentucky that year.[148] The census figures, released in February 2009, were somewhat surprising because there were a couple thousand more tobacco farms than had been estimated. The remaining number of small tobacco farms was even more surprising. I was told many times—by former and current tobacco farmers, extension agents, and others—that tobacco farms are all becoming large farms. Tobacco farms are becoming "industrial" farms, I was occasionally told by retired and former tobacco farmers. However, according to the Census of Agriculture, in 2007

Kentucky and North Carolina Farms Growing Tobacco, 2007				
	Kentucky		**North Carolina**	
Tobacco Acreage per Farm	**Number of Farms**	**Percentage of Farms**	**Number of Farms**	**Percentage of Farms**
Less than 2 acres	957	12	92	4
2.0 to 4.9 acres	2,688	33	212	8
5.0 to 9.9 acres	1,991	25	235	9
10.0 to 24.9 acres	1,687	21	503	19
25.0 to 49.9 acres	497	6	443	17
50.0 to 99.9 acres	221	3	554	21
100.0 to 249.9 acres	69	<1	487	19
250.0 acres or more	3	<1	96	4
Total	8,113	*	2,622	*
Source: US Census of Agriculture, 2007, Bureau of the Census, National Agriculture Statistics Service, USDA (released in February 2009) * Totals will not equal 100 percent, due to rounding.				

the largest group of Kentucky tobacco growers (33 percent) raised between 2.0 and 4.9 acres; only 10 percent raised 25.0 or more.[149] This is even more startling when compared with North Carolina, which had just over 2,600 tobacco farms in 2007, with the largest category in the over-one-hundred-acre range (583 farms).[150] A persistent thread throughout this book is an attempt to understand the discrepancies between public perceptions of the number of tobacco farmers and census figures, as well as conflicting perspectives on the future of Kentucky tobacco.

The following exchange took place in early 2008, during an interview with Jonathan Shell and his grandfather, G. B., with whom I had visited many times over the previous year:

Author: I was asking your granddad what he thought about me coming down and learning from ya'll, for really almost a year—

Jonathan: Well I like it. I hope you romanticize tobacco in your [book] and you get published; that way people will start loving it.

Author: So you think I should romanticize it?
Jonathan: Yeah.
Author: And what does that mean?
Jonathan: Just make it intimate. To where that they can see that you know, there's hands that touch this stuff, and that there's lives that are dependent on it.

My goal in this book has not been to romanticize tobacco farming, but I do hope that I have succeeded in demonstrating that "there's hands that touch this stuff, and that there's lives that are dependent on it."

Part 1

The Burley Tobacco Crop Year, Then and Now

Introduction to Part 1

It's a late winter day in March 2007, and when Keenan Bishop, county extension agent, and I get out of my pickup truck at a Franklin County, Kentucky, farm, we are immediately encircled by two curious but friendly dogs, including a tiny dog I will come to know as Buster. We walk through a gate to where two men are working, and I meet Martin Henson for the first time, along with an older fellow helping him with the day's work. As we talk, Martin continues to work, and following Keenan's lead, I begin to help a little. What I didn't know at the time is that Martin would become central to my fieldwork and that the pattern of interaction with Martin, a pattern I would follow over the next year, was being set.

In part 1 of this book, I illustrate the tobacco year, describing the steps in the process of raising and selling a crop of burley tobacco, based on what I learned during my fieldwork in central Kentucky. My fieldwork began in 2005, but I spent the most time in the field as I followed the 2007 crop. I conducted recorded interviews and spent time with farmers (male and female), retired farmers, wives of farmers working off-farm jobs, county agricultural extension agents, university tobacco specialists, former and current warehousemen, and agricultural organization leaders. I was treated to numerous farm tours, which usually included inspection of the tobacco in whatever stage it was in at the time and often a comparative inspection of tobacco being raised on different farms by the same farmer; tobacco barns and stripping rooms; greenhouses, float beds, and plant beds; and sometimes equipment sheds. Many such tours included only the parts of the farm directly relevant to tobacco and not the spaces in which cattle grazed or were fed or where corn or other crops grew. I was invited to visit and observe—and at times participate—at crucial steps in tobacco production and marketing; attended events organized by farm organizations; and was generally welcomed into the farms, homes, and offices of many generous people. I was frequently fed dinner at midday and sometimes supper in the evening;

the location depended on the role of women in the family as well as the intimacy of the relationship, with meals either in farm kitchens or at nearby convenience stores or restaurants.

I later realized that Martin Henson never gave me a farm tour. Instead, he served as my guide through the tobacco year—even as I worked with many others in the tobacco community—and so I saw his float beds as he prepared them and his stripping room when it was time to strip tobacco and so on.[1] I also accompanied him as he fed his cattle, if I was there when it was time to feed the cattle. Martin welcomed me into the daily routine of the farm in a manner that allowed me to learn the details of raising a crop of tobacco as I needed to learn them but also to experience the rhythm of work on a tobacco farm, which most often also includes at the very least activities related to cattle, corn, and hay. Because Martin was a central figure in my learning process, he is also central to the following narrative I have constructed about the crop year.

That said, however, Martin's experiences and perspectives cannot be taken to represent the experiences and perspectives of anyone but Martin. One result of the many changes that have taken place in tobacco production and marketing is that, as tobacco grower Roger Quarles pointed out to me, it is now more difficult to describe a *typical* burley tobacco farmer. At one time, although there were certainly differences based on economic class and farm/land ownership, farmers largely faced the same issues, and their practices varied only marginally from farm to farm or county to county. Crop sizes certainly varied, but not widely as compared with today's variations. The current period of transition has brought with it much more variability, in part because acreage size directly relates to farm practices. Over the course of my fieldwork I met the spectrum of growers, from small growers who continue to do the bulk of the work themselves to growers of large acreages who, as one farmer described it, "never touch a stalk of tobacco."

As retired tobacco farmer Jerry Bond told me, "Tobacco's commonly known as the 'thirteen-month crop,' because, most times you're still finishing one crop while you're starting the next year's crop—preparing the soil or the seed beds or whatever. You know you may be stripping tobacco, when you're starting the next one." The chapters in part 1 follow the thirteen-month production cycle, highlighting changes that farmers described to me. During interviews, casual conversations, and situations such as farm tours, I was told not just about how tobacco work is done today. I was also told about how the work used to be done, about the changes that are currently taking place, and about the changes that could be seen coming down the pike. A prevalent stereotype about farmers is that they are resistant to change. This is a particularly widespread belief about tobacco farmers, because as the economic and political/social context of tobacco production has changed, public discourses have continually asked, "Why don't they just grow something else?" There are a host of reasons why just growing something else is not a solution for many tobacco farmers, and I will discuss some of these in this book. Here it is important to point out that many assume that tobacco farmers are simply unwilling to experiment with new ideas. The many changes that farmers have experienced that I describe in part 1 of this book—including many that farmers initiated—demonstrate that tobacco farmers are very open to new ideas, once it is clear that those ideas make sound economic sense in their specific circumstances.

Chapter 1

Sowing the Seeds and Setting the Tobacco

The work that Martin was engaged in on that March day when I first met him was the preparation of the beds in which his tobacco plants would germinate and grow. Because tobacco seeds are too delicate to plant directly in the field, they must be started in a protected environment and then transplanted. As we talked, he stretched thick black plastic across wooden frames built directly on the ground, weighted it down, and filled the plastic-lined frames with water. Later, he would plant seeds in polystyrene trays filled with peat-moss-based soil, and the trays—each about thirteen by twenty-six inches, with about 250 cells[1]—would then be set in the water to float in these *float beds.* Fertilizer would be added to the water and the beds covered at night and uncovered during the warm days of spring.

Before the 1990s, all growers started their tobacco plants directly in the ground in one-hundred-foot-by-twelve-foot plant beds. The preparation of the plant beds was a major task that evolved over the years but retained the same basic parameters: ground was cleared, weeds were killed, seeds were planted and covered, the beds were weeded, and the plants were individually pulled and transplanted into neat rows in the field. Since at least colonial times, through the 1950s and '60s, tobacco beds were "burned," which meant that burning wood or brush in some form was used to kill the weeds. Early on, burning logs were slowly rolled across the beds, and later, fires were built on metal frames that were then dragged across the beds, resting for a period on every part of the bed. *Burning the beds* was an event that took place either in the late fall or in the early spring, depending on the practices of a particular farmer. For the children in some farm families, burning the beds was an annual event to look forward to, as it meant staying up late to monitor and move the fires, perhaps roasting hot dogs and marshmallows. In the 1960s, it became common practice to *gas* the beds—the prepared plants beds were covered with plastic and gassed with a range of chemicals, the most popular and lasting of which was methyl bromide, which was sold in pressurized cans. Once the beds were treated for weeds, they were covered to protect the seeds as they germinated; they were fertilized and irrigated as they grew. Whether burned or gassed, some weeds would survive, and the task of weeding the plant beds without stepping on or otherwise harming the young tobacco plants was a dreaded job.

When it came time to *set* tobacco—as transplanting the young plants into the field is called—individual plants were pulled from these plant beds and brought to the field, either in baskets or wooden crates or wrapped in burlap bundles. Women often *pulled plants* as men prepared the ground for setting or started setting—one woman commented to me that it seemed like whenever there were plants to be pulled the men suddenly had ground that needed to be worked—but many people told me it was a job that "everyone" did. Pulling plants is often described as one of the worst aspects of tobacco work because it meant being bent over for long periods of time, trying to get through the bed without stepping on plants, which often meant perching precariously on a wooden board that was balanced across the bed. During a 2005 interview, Kathleen Bond described it vividly: "It's hot and you're like, either standing on your head or squatting down and, you know, neither one is comfortable and you have to—you can't like really walk out into the bed because you, you know, step on plants and ruin them and you have to—I don't know, it just, it makes you sore and the sweat's running in your eyes and deer flies get on your back and bite you." By the late 1990s, the vast majority of farmers moved from plant beds to *water beds* or *float beds*.

The practices involved in starting the plants from seed provide one example of my statement that there is less typicality today. Many tobacco farmers have built greenhouses that house their float beds. Greenhouse float beds produce more usable plants. Robert Pearce of the University of Kentucky told me that according to results of university trials, greenhouse float beds yield about 95 percent usable plants, versus 80–85 percent in outside float beds. Martin estimated that he sees about an 85 percent success rate in his outside

By the late 1990s, the majority of growers raised their tobacco plants in float beds—with polystyrene trays floating in water—rather than in the ground. (Photo by author)

float beds. Although greenhouse float beds may have a higher success rate under ideal conditions, greenhouses are more expensive to build and maintain than outside float beds, and they introduce additional risks. Greenhouses provide opportunities for controlling the environment in which the plants grow, particularly the temperature, but heating and cooling systems must be carefully monitored in order to avoid temperature extremes. Greenhouses can also harbor diseases that can spread quickly through an entire greenhouse and possibly wipe out all the plants in it. Growers that I met had experienced such disasters firsthand. For all of these reasons, some growers have chosen not to grow their own plants and instead buy them from other producers. In one sense, this can be seen as a promising entrepreneurial opportunity for farmers as they adjust to the decline of tobacco markets, and many look for additional or new sources of income. On the other hand, the buying and selling of tobacco plants has been described to me as a practice that represents not only the changes that have taken place in tobacco production but the loss of

Many larger growers have built greenhouses in which to start their tobacco plants in float beds. (Photo by author)

the community both maintained and symbolized by reciprocal or swapping relationships between farmers.

However, because of the difficulty of the work involved in pulling plants, the transition to float beds was quite welcome. Martin told me this about his transition:

> When I first started out, I was gonna try it, so I put out about a hundred trays. And at that time my wife, who's a nurse now [*slight laugh*], she was, that was before she was going to school. And this is back in the '90s, early '90s. I had plant beds and then I had these hundred trays, on water. Well, we pulled the plants, and then, when we got them done, we went to the water beds and, used them. She said "I'll never pull another plant" [*laugh*]. So I had to [*laughing*] make my water beds bigger, because I sure couldn't pull them.

The next year Martin planted all his plants in float beds. I've heard

Prior to the 1990s, all growers started their tobacco plants in one-hundred-foot-by-twelve-foot plant beds. The Waits family continues to raise half of their tobacco plants in plant beds, as shown here. (Photo by author)

similar stories of rapid movement to float beds from other farmers. But even as I write about this in the past tense—because plant beds are talked about almost entirely in the past tense, as a practice of the distant past—I came to know one family who continues to raise half of their plants in plant beds and half in water beds. Marlon Waits, who raises burley tobacco with his brother and cousin, told me that they consistently see higher yields from plants started in the ground; plants beds "just seem to make a better plant," he told me. I was told about other farmers who also continue to start some or all of their plants in the ground.

In addition to relieving the farmer from having to take the time or suffer the misery of pulling plants, the advantage of float beds is that fewer plants are wasted. When plants had to be pulled from plant beds, the morning was usually spent pulling plants, and the afternoon was spent setting them in the field. Each day, guesses had to be made about how many plants could be transplanted into the field that afternoon; few would survive for tomorrow if too many were pulled or if a storm came up. With the float bed method, the trays full of plants are simply pulled out of the water and loaded onto a wagon or truck still in the tray; any leftovers can be slipped back into the water.

Although seed is costly, it is common practice to sow more than a grower will need in preparation for the possible loss of plants due to disease or harsh weather conditions such as drought, frost, or hail. This was particularly true when plant beds were the norm, since plants were also lost if they were pulled and not used. If all went well, there would be plants left over after the crop had been successfully set in the field, which meant that if a neighbor ran into problems and needed more plants, surplus plants could be given to the farmer in need—literally a life-saving act when tobacco was the only source of cash for a farm family. According to Roger Quarles, who raises extra plants and sells them, "Customarily everybody put out their own tobacco beds and then if they, if for some reason theirs didn't do well or had bad luck or something—typically you'd go across to your neighbor, maybe he had some extra or something and there just, never was any charge." He went on, "Occasionally you'd hear something like that, somebody charging for plants. And if they did they'd be talked about [*slight laugh*]. But then it became apparent that people didn't mind paying for them so it was a very quick shift."

Robert Pearce, extension tobacco specialist in plant and soil sciences at the University of Kentucky College of Agriculture, told me:

> It's kinda interesting—and this is just I guess sort of a social aside—is that before, when you had plants in beds, you'd go and you'd pull your plants and when you got done, you'd have extras—we seeded a lot more plants than we ever used in those days. And so, when you got done and a neighbor called up and said "I need some plants," yeah, you know they'd just come over and pull them. You know, we didn't really think about plants having a value. Of themselves. And now the culture is that with these, with the greenhouse plants, they have a very real value and can be bought and sold and traded.

As Pearce, Quarles, and others have told me, float beds have dramatically changed the value of young tobacco plants, making them a commodity to be bought and sold. That farmer in need may now have to buy those plants from neighbors or from growers who can be located through the University of Kentucky College of Agriculture website. This change is seen by some as symbolic of the loss of the swapping relationships between farmers. Although this change may also signify a larger shift in communities more generally, it is seen by some tobacco farming families as a negative consequence of the move from plant beds to float beds even as they are thankful for the benefits of float beds.

Martin raised all his plants in the beds he was preparing that first day, enough to plant just over thirty thousand pounds of cured leaf (approximately fifteen acres), spread over three farms—the amount that he had raised for many years. When I say that Keenan and I "helped" on that first day, I mean small tasks like picking up a couple of pieces of wood from the back of Martin's truck to weight down the edge of the float bed liner and helping to stretch out the liner while Martin filled the beds with water. I later realized that what I *didn't* do was important; I didn't just stand and watch, distracting him from his work, expecting him to stop working while he talked to me. Instead, Keenan and I slipped into the rhythm of Martin's work, assisting in a minor way without being asked or asking. In two different contexts, Martin told me a story about him and his wife, Kathy's, first date, which they spent horseback riding on the farm. This version was recorded in a storytelling session that developed when Martin came back to the house as I was finishing an interview with Kathy: "She came out one evening, Sunday evening I believe it was or something, but anyway I loaded up some saddles and we went back there and the horses are back in the back. We got to this gate down here and, I pulled up there and, she sat there and, I got out, opened the gate, got back in the truck, pulled it up, got out, closed the gate, got back in the truck. I said, 'Now girl, if you're gonna spend any time around here, whosever driving don't have to open and close the gate.'" At this point Kathy, who was not raised on a farm (and who, after several years of working the farm with Martin once they were married, went back to school and became a nurse), said, laughing hard, "I just sat there. I didn't know I was supposed to open the gate!" This brief narrative exemplifies Martin's expectation that anyone spending time with him on the farm needs to be able to anticipate basic tasks that need doing.

During my many visits to Martin's farm I assisted in small ways when I could, and he taught me to do some of the tobacco work, including setting, topping, and stripping. Hanging around with Martin as he worked meant learning to anticipate some small task—like opening a gate or hitching a trailer to his truck—that needed doing. Usually, not being a farm girl, I had never done these things, and I did my best at the tasks, hoping most of all to avoid embarrassment. Occasionally, however, I would have to ask him how to do something, and there went the rhythm of work and talk as Martin had to show me how to do whatever it was, always something that was second nature to him. In January 2008, as we pulled up to the barn in his pickup, just having gone through the main gate leading onto the farm, he again told me the story about Kathy and the gate—but he prefaced it with, "I'll tell you like I told Kathy . . ." Not understanding why he was telling me this story again, and why he was prefacing it in this way, I protested, "But I opened the gate!" and then immediately I realized that for some reason, on this occasion, I had asked as we pulled up to the gate, "Do you want me to open the gate?" instead of routinely opening it.

At other times, the rhythm of work and talk with Martin meant standing on the back of the tractor or the highboy as he drove back and forth through the rows, pulling the *setter,* spraying *sucker control,* or dropping tobacco sticks. It sometimes meant riding along with him in his truck while he fed cattle, often with Buster asleep in my lap. On a couple of occasions, it meant helping his friend Elic, a farmer with whom Martin has a close reciprocal work relationship. And it meant topping and stripping tobacco beside him and his crew.

As he set up his float beds that first day, Martin, sixty-six years old at the time, described his situation as a tenant farmer on the farm where we were that day. He has rented this farm, about 275 acres, for over thirty years for part of his tobacco crop and for his cattle, hay, and corn. There are only about thirty tillable acres on this farm—typical of the hilly farmland of central Kentucky, explaining why tobacco and cattle have long been an important complementary pairing in this region. As Lincoln County extension agent Dan Grigson explained in an interview, "Of course, Kentucky is more suited to forage and livestock than anything else. You have the rolling land, it's not suited for row crops, you're not gonna grow many corn and soybeans on a lot of the land that we have, especially in central Kentucky and in east Kentucky." Henry County tobacco farmer Mark

Roberts told me, "We don't have those big, huge thousand acres that we can go out here and rent and put in corn and go over it in a few days and be through." Historically, this labor-intensive crop was raised in very small acreages with growing quotas that averaged less than five acres at one time, and quotas of a half acre or even less were not uncommon. As I have described, the quota system changed to pounds in the early 1970s. Even today, tobacco remains the most profitable small-acreage crop for many farmers, and although acreages are increasing, the average crop size was 10.8 acres in 2007 (up from the post-buyout average of 3.8 acres), and on 45 percent of Kentucky tobacco farms fewer than 5 acres were grown.[2]

Martin's main farm, where we were on that first day, is where his barns, equipment sheds, and tobacco stripping room are located, but he also rented land for his tobacco and cattle on another farm nearby. He had purchased about eight acres of tillable land with his buyout money, on which to grow about half his tobacco crop. He and Kathy had once lived in the old farmhouse on the main farm, but a few years ago they bought a small piece of the farm and built a house that we could see from where we stood. One day, Martin and I broke for lunch after a morning of setting tobacco and joined other farmers and local men who regularly frequent the local convenience store and gas station at lunchtime. The men joked with one another, told stories on each other, and talked about their crops and livestock. Many of their stories and jokes featured Martin. Although this may have been for my benefit, Martin is both well liked and full of stories, so I imagine stories told by and about him often take center stage. At one point, Martin told us a story about the owner of his farm offering to sell it to him for about $1.7 million; he and the other men just laughed and laughed. Martin said that the owner had told him that although it was a lot of money, he could hold on to it, and someday it would be worth even more. Martin told him, "Yeah, *you* hold on to it and it will be worth a lot too!" As the men laughed at the landowner who thought that Martin could come up with $1.7 million or would be willing to take on that kind of debt, they were also commenting on the rapidly rising land costs of central Kentucky, rising costs that make ownership of a farm impossible for a farmer who doesn't inherit one.

For many, "tenant farmer" connotes particular images, perhaps negative, of a time long past. Lu Ann Jones points to the tendency of scholars to "dichotomize southern farmers" into the poorest sharecroppers and "planter-landlords who manage other people's labor," when in fact a "mosaic of tenure arrangements characterized southern agriculture, and farmers might belong to more than one group simultaneously."[3] Tenant farming has remained a common practice in tobacco regions. Martin's situation is typical; most farmers who raise more than a very small tobacco crop piece together their total acreage over multiple farms in order to maximize their tobacco income. This is due in part to the hilly Kentucky landscape and in part to the practice of raising tobacco on your best land, but it is also largely due to the quota system of the former federal tobacco program.

TOBACCO SETTING TIME

May 10 is commonly cited as the earliest date to begin transplanting or setting tobacco because the chance of frost will finally be past. Others start much later; Martin told me he was raised to set tobacco on Memorial Day and *house* it on Labor Day. Holidays often serve as markers for tobacco work, as people also talk about stripping tobacco on Thanksgiving and the ideal of having it sold by Christmas. Marlon Waits told me that they were once always weeding tobacco beds on Kentucky Derby Day, and that the Fourth of July and Christmas were the only holidays around which he could make plans that didn't involve tobacco. Jerry Bond described the associations between tobacco and holidays explicitly, but he also had different things to say about some of these same holidays:

> Tobacco is a holiday crop. Whenever the holidays come around you're doing something in tobacco. You're always setting tobacco on Memorial Day. You're always chopping out tobacco on the Fourth of July, chopping the weeds out with a hoe. You're always cutting tobacco on Labor Day. You're always stripping tobacco on Thanksgiving, and sometimes on Christmas. I mean a lot of times we would eat a holiday meal and then go to the stripping room, or go to the stripping room and then come and eat a holiday meal. You know you mix work and holidays. So it's associated with every holiday.

This reflects both tobacco's status as a "thirteen-month crop" and its symbolic connections with all aspects of life, mundane and sacred. I was told more than once about wedding dates set with the tobacco calendar as the guiding factor.

Before setting can begin, the ground must be *worked*—plowed, disked, and otherwise prepared, dependent on the weather and the practices of a particular farmer. In order for the ground to be worked, there must be moisture.[4] In May 2007, it was still too dry for Martin to set his crop. It had not rained a drop on him in May, and he'd already mowed the plants off twice and would have to do so again soon. Using various technologies, tobacco growers may trim their young plants back multiple times before they are set, a practice that started out as a means of holding them from growing too large if setting were delayed by weather; early attempts involved the use of a string trimmer (or weed whacker). It has since become a management practice that is used to strengthen the stems and to produce a more uniform and generally healthier crop, as it evens out the height of the plants and ensures adequate air circulation among them. This practice is one of many that demonstrate the interaction between tobacco farmers and university research. According to tobacco specialist Robert Pearce, farmers came up with this practice in order to hold their plants for late setting, as I describe, and the University of Kentucky then conducted research that led to the current management practice. In Martin's case, mowing off the plants meant taking each tray out of the water and passing it under a lawn mower attached to a stable base and then putting it back in the water; this requires about four people, difficult for Martin to find this time of year. Those with greenhouses have systems that allow them to mow plants without removing the trays from the water. This usually involves a modified lawn mower that is suspended over the plants and pushed across with relative ease.

The spring and summer of 2007 was a time of severe drought, a drought so severe that the closest point of comparison for farmers and extension agents alike was the early 1930s—a drought that few are old enough to remember but most had heard stories about. If they had not heard the stories before, they heard them in 2007. Various years were given to me for this drought, but according to the Kentucky Climate Center at Western Kentucky University, the worst drought to hit Kentucky was in 1930–1931.[5] The other point of comparison that Martin and others mentioned was the drought of 1983, a drought that until 2007 had been the worst in the memory of many. At the end of the crop year, it turned out that 2007, although dryer than 1983, was not as devastating for most tobacco growers. Where farmers really felt the effects of the 2007 drought was in their cattle operations, as they faced severe shortages of hay and other grains, and many were forced to sell their cattle early when faced with the expense of feeding them through the winter.

As farmers began telling me around July, tobacco is a "desert crop" that needs only a small amount of rain, but at specific times—after it is set, after it is topped, and before it is cut. Martin told me that an inch a week is ideal. Clarence Gallagher of Fleming County explained to me in late June 2007 that "tobacco is a deep, kind of a deep-rooted thing" and that "dry weather is actually good on tobacco. If you get the right rains at the right times." He went on to explain that the roots of the plant will be forced to hunt for water in dry weather and that

> when you get that many roots out there, that plant is gathering up everything that you put on that ground. You know when you put your fertilizer on you've got more roots there, to take in whatever you've got on the ground. And really, in wet weather? It'll set right on top of the ground. And grow. And I mean it won't, it don't have to fight to get what it needs. I mean it'll actually just set right on top of the ground just wherever you can set the root at. It won't have that deep tap root that'll go down there and try to find [water]. And I actually like to see dry weather when I first set, just to get a good root system, and then you know let some rains come and then it's got the roots and everything out there, to grab whatever falls out of the sky then and it's, you'll have a whole lot better tobacco crop.

I was told by several farmers that a wet crop will starve you to death, and a dry crop will scare you to death. As the months went on, and we moved into the fall curing season, I began to be told that the most important time for moisture is the fall. Without moisture tobacco will not cure; it will simply dry up. That's what happened to a lot of the 2007 crop, as I will describe.

It was mid-June before Martin set his crop. When I arrived one morning that June, he and his crew of four workers had been at it since about 7 a.m., and when I got there they happened to be stopped, loading the setter with trays of plants. We exchanged greetings, and once they were loaded up, Martin drove off into the field on the tractor, pulling the setter and his crew. This was a fairly large field with a hill sloping down and away from where I stood, and so I watched and waited for what seemed like an eternity as they disappeared over the hill and then finally came back into sight. This was only my third visit to Martin's, and since this was the setting of the crop, it was the first time he was on the tractor when I arrived. As I waited, I wondered if I'd be spending the day watching the tractor come and go, and I asked myself how long I needed to stay here before I could find a way to politely excuse myself, if this was going to be the extent of it. But when he came back and turned the setter around to enter the next set of rows, Martin stopped and asked if I wanted to ride behind him. I spent the morning standing on the hitch between the setter and the tractor, holding on to the fenders, leaned over toward him so that I could hear the stories or pieces of information that Martin volunteered and his responses to my questions. At one point, he asked if I wanted to set some plants. After I had set a few rows, and I was getting ready to resume my position behind him on the tractor, he asked me if I was left-handed or right-handed. When I replied that I was right-handed, he said, with a slight gleam in his eye, "Well I probably should have told you to sit on the other side of the setter then. It might have been easier."

The process of setting tobacco changed considerably through mechanization at about midcentury and has undergone only relatively minor changes since.[6] The older farmers I've talked to described setting tobacco by hand as children. They set it *by season,* meaning that they waited for a rain so that the ground would be soft, and then they dug each hole with a finger or a specially made peg that was whittled from a stick or broom handle, or in some cases a metal peg that was purchased, and dropped the plant in and closed the soil in around it. In the 1920s, a new innovation for setting tobacco was introduced, a small metal device with a pointed end and a large compartment that held two or three gallons of water and a smaller compartment into which the plant was dropped. Still a hand-setting method that meant planting one plant at a time, this small device with various names, including *jobber,* eliminated the need to poke the hole with a finger or peg and lessened the need to set by season.

The first mechanical setters were also developed in the 1920s, but many farmers were unable to purchase one until as late as the

Martin Henson setting tobacco. (Photo by author)

1950s. The early setters were of course pulled by horses. Many of the farmers I talked with who are old enough to remember the move to tractors saw them introduced as late as the 1950s and 1960s. Early tobacco setters were one-row setters that required two people sitting on seats that were barely off the ground and one person to drive the team or tractor that pulled the setter. The two people on the setter took turns putting the plants in the ground as the setter opened and closed a furrow and watered the plant. Tractor-pulled two-row setters are now common, and they require a driver and two or four people riding and setting, depending on the design. Both women and men ride the setter, but driving the tractor has traditionally been a universally male job—so much so that I was told several stories about wives who wanted to drive the tractor and pull the setter, and when they finally did they only caused trouble. According to Diana Taylor, her father and other men drove the tractor during setting time not only because the tractor took physical strength to control but because the straightness of the rows—a public performance written on the landscape—is a point of male pride. According to Eric Ramírez-Ferrero, "evaluations" of how a farm looks serve as

A typical four-person tobacco setter. (Photo by author)

"physical manifestations of good decision making" and pride.[7] Pete Daniel has written that "farmers always associated crooked rows with sorry people."[8]

When setting tobacco with a typical setter, each plant is placed by hand into the mechanism of the setter; depending on the design this may be metal fingers or a wire basket. The machine does the rest—it digs a furrow, adds a water/fertilizer mix that the grower has prepared in a tank attached to the setter, drops in the plant, and covers the furrow. This all happens very quickly, and it requires constant movement and coordination on the part of whoever is doing the setting: pulling a plant from a tray with one hand, possibly shifting it to the other hand, and then dropping the plant into the rapidly turning mechanism. With a finger-type setter, the hand that is dropping the plant must wait to be sure that the mechanized fingers grab the plant at just the right point before reaching for another plant.

Plants are occasionally "missed," and following the setter to fill the gaps is a job described to me repeatedly as the first responsibility in the tobacco field that many adults remember being given as very small children. After the adoption of the mechanical setter, jobbers were often used for this task. Prior to 1971, tobacco quotas were allotted by acre—or by a percentage of an acre—rather than by poundage. Under this system, growers worked to raise the most tobacco they could on their allotted acreage since whatever could be raised on a farm's allotted acreage was later sold by the pound, making every leaf valuable. This also meant that each year, government officials went to every farm on which tobacco was raised and measured what had been planted, destroying any part of a crop that exceeded that grower's allotment. Today's growers rarely expend the labor needed to set missed plants by hand; now that farmers are not limited by an allotment based on acreage, larger crops are raised, and most of the work is done by paid laborers instead of family members.

Already late because of the dry spring, Martin's crop got even later when his tobacco died just after he set it. He had to reset almost every row of every acre. The field that I'd "helped" (as he likes to say, despite my helping very little) set was completely disked under and reset, while in his other fields there were a few rows that survived. It was nearly July 4 before he had a healthy crop of tobacco set, and there was still no rain. This was the latest he'd ever set tobacco in his nearly sixty years of raising the crop, and it was the most he'd ever had to reset. In February 2008, he was still talking about what he should have done differently. He told me, "I was uneasy about it when I set, because the ground didn't suit me." He told me he should have irrigated more before he set so that the ground would be more welcoming, and he could have worked the clumps out of the soil more thoroughly. This might have avoided the loss of that crop, a loss that amounted to over $5,000 in labor, plants, and chemicals—even with about 250 trays of plants given to him by a neighbor (despite the concerns of many, reciprocal relationships among farmers still do indeed exist).

Discussions of what he's done with his crop, what he plans to do, what he should do, what other farmers have done or plan to do—all these things are part of the daily life of the farmer, whether leaning on the pickup of another farmer who has stopped by to borrow equipment or lend a hand or at midday, among those who meet up at a local establishment for lunch. I was told repeatedly that no crop of tobacco is like the one before it or the one that will follow and that for many, the challenges that each new year brings are part of the appeal of raising tobacco.

Tending the Crop

Certain steps in the process of raising a crop of tobacco have come to symbolize tobacco work, and these practices are discussed more frequently than others. This includes pulling plants (even though this practice has become rare), setting, topping, cutting and housing, and stripping tobacco. Perhaps these tasks are talked about most often because they were times when the whole family came together in the work, and they therefore are the times about which nostalgia is most often expressed. Other tasks—side-dressing, the application of herbicides and pesticides, cultivating, and *chopping out,* for instance—are mentioned or described with less frequency. Often they were not mentioned to me at all unless I asked about them. These are the tasks that are most frequently performed by the lone farmer. The rows are side-dressed (fertilizer applied) after setting, cultivated after setting, and the weeds are chopped out periodically throughout the season, depending on the conditions of the season and the amount of time available to the grower.

The time from tobacco setting to when it is cut and housed spans about ninety days. After tobacco is set, cultivated, and side-dressed, there is a bit of a lull in the tobacco work during which a grower can turn attention to cattle and other farm tasks. Most tobacco farmers also raise cattle, and when I asked Martin why they go together so well, he responded:

> Well, when you're, except for stripping, when you're in your tobacco, there's not a whole lot to do with the cattle. In the spring you work them. Of course I've got cows calving in the spring. You work your cattle, you turn them out, you look up at them two or three times a week or whatever. And you're in your tobacco, and cutting your hay. Of course you've got to cut your hay and all that. What I like to do, I like to cultivate tobacco, early in the morning, and then go cut hay, and then, do that and then the next day, cultivate a little tobacco and go tend the hay. And then, the next day, cultivate a little tobacco and go rake hay and bale. That way it's kind of a one-man operation.

He went on, "But, cattle and tobacco work out pretty good for me. When you're slack on cattle, you're you know busy on your, cutting hay and plowing, and chopping out and so forth." In good years, the cattle need very little attention, but there is hay to cut and bale, corn to harvest, and silage to chop, all in preparation for the needs of the cattle in the coming winter. In drought years things are different. In 2007, farmers were beginning to feed hay to their cattle in June because of the lack of forage for grazing, and there was very little hay to cut.

At some point, however, the weeds must be chopped out of the tobacco with a hoe. The number of times varies, and the degree has changed dramatically over the years due to the introduction of herbicides as well as increased acreages. Charles Long, in his late sixties when I interviewed him in 2005, described chopping out when he was young: "They were particular about it from the time it went in the ground until the time it went on the tobacco floor. Used to they raised tobacco, they called it 'choppin' it out.' You went through with a hoe, and you," he went on, "cut all the weeds out." He told me that he worked for a guy for whom "it didn't make any difference, if there wasn't a sprig of grass or a weed, [with]in ten feet of that plant, he wanted the dirt pulled away from that plant, and new dirt pulled to it."

Several times, I was given descriptions of tobacco that was full of weeds because a farmer just didn't take the time he should have to properly chop it out. One extension agent told me that there are farmers today who don't "care about the weeds in it as long as there's not so many weeds that they can't get the cutters through." The contrast between this and Charles's description makes it clear that the standards are understood to have changed, and this is a point of judgment for some.

In 2007, the severe drought conditions lessened the need for chopping out but made it necessary to irrigate, for those with the resources to do so. Irrigation means a lot of work moving irrigation equipment between fields, and it often means late nights of watching the pump. I went out to visit with Martin one evening while he irrigated a field at the back of the farm with water from a small lake nearby. He connects a pump to his tractor, which then connects to a complex system of pipes and water guns; the pipes and guns must be moved as he irrigates the field in pieces over several days. He described for me the complicated process of setting up to irrigate in one of our interviews, laughing as he said, "And then, you go down and babysit the pump for, three hours or four hours, however long it takes, and that's, that's worse than work."

Tobacco is closely observed by the grower—and his or her neighbors—during this period of time. Clarence Gallagher and I walked out into one of his fields one day in June and discovered tobacco worms on several plants, even though he had already sprayed to prevent them. He remarked during an interview later that day, "If I'd not walked out there for two weeks they'd have eaten it all"—but of course he would never have left his tobacco that long without inspecting it. Walking out into the tobacco regularly is as important as any other part of tobacco production, as the grower vigilantly watches for signs of disease or pests. Looking at the tobacco was for many a part of family ritual, according to extension agent Keenan Bishop:

> Keenan: I think it was traditional. We'd go there on Sundays for dinner, and you know, if it was during the summertime, part of that ritual was you'd get in the car and drive out in the field and, look at the tobacco and, "oo" and "ah" and, it was my grandfather's way of showing off and, it was just kind of an expected thing. You knew if tobacco was growing you had to go look at it because it was, you know it was so important, the primary part of their life.
>
> Author: And would he point out, that it looks good or it looks bad or that—
>
> Keenan: Oh yeah you're always measuring, you know the height of it and how it compares to neighbors' and, how it compares to last year's and, the size of the leaves and all that stuff and, you look through the family albums and you see, you know pictures of kids' birthdays and Christmases and tobacco crops.

Phyllis Bailey's *Shelby County Tobacco Farmers: A Pictorial History* demonstrates this, as it includes not only photographs from throughout the tobacco year but many photographs of families standing in or in front of their tobacco patches. Although the majority of the other photographs in the book are snapshots of tobacco work as it was being carried out, these photographs are all clearly posed.

Before the advent of pesticides, tobacco horn worms and other pests were picked off by hand—another series of walks through the field, this time picking off these chubby green worms (thick as your thumb), squishing their heads, and dropping them into a bucket. There are many diseases to look for, but the most destructive diseases are blue mold and black shank, which, despite the availability of resistant varieties, frequently show up. Blue mold is airborne and is bad in wet years; it has often been spread by tobacco plants grown elsewhere and trucked into the region. Black shank can be waterborne and therefore can be spread by irrigating tobacco with infected pond water; once it is in a piece of land it stays there. Farmers often talked about there being more diseases to be concerned with now than there once were. Dan Grigson told me, "It's not a crop that you just put out there and let it grow. It's something that [my dad] looked at *every day,* when that crop was in the field he would check it every day. You didn't drive by in the pickup truck either, I mean he would get out and walk through the crop and see what was going on. If there were disease or bugs or weeds in there then they had to go, you know." So although this time period—usually June and July—is often described as a slow period in terms of a farmer's tobacco crop, there is plenty to do.

The lull ends when it is time to *top* tobacco—to walk through the fields and snap out the bloom at the top of every plant by hand. Flue-cured tobacco producers mechanized this process in the late 1960s,[9] but it is said that burley tobacco is less suited for mechanization of all kinds, including topping. Burley farmers—whether they raise three acres or three hundred—believe it takes a human hand to snap out the bloom in just the right place so that the flower is fully removed and no leaves are ripped or wasted. According to extension professor emeritus George Duncan, although burley farmers would not accept the mechanization of topping because they felt they were losing the high-priced top leaves, the labor they would save through mechanization might make up for the leaf loss.

Topping usually takes place in July or early August and is an extremely hot and sticky job. In a normal year, the tobacco will be well over the heads of those doing the topping. In 2007, I topped in a field of tobacco that was between the height of my knees and my shoulders, much of it requiring bending rather than reaching (but it was still hot and sticky). It's a hard job: walking through and alternating between two rows, snapping out the bloom or just the right number of top leaves if a bloom has not yet fully formed, without ripping the larger leaves around it. It's easiest when a bloom has started and when it's a tall plant. When I topped with Martin and his crew, I topped just one row at a time while they each topped two, and I still quickly fell behind.

Topping also signals that it will be three to five weeks until cutting and housing can begin. Around topping time, maleic hydrazide (MH-30) is applied. MH-30, which goes by a number of vernacular names—*sucker control, sucker dope, sucker kill,* or simply *MH*—is a chemical that stops the growth of the "suckers" that sprout in the elbowlike space between the stalk and each stem in order to encourage the growth of the valuable leaves of the tobacco plant. Before the introduction of MH beginning in the 1960s, all the suckers, like the tops, had to be broken out by hand. *Suckering* tobacco once meant walking through the fields, multiple times, removing the suckers at every joint of every stalk of tobacco. According to Bob Taylor, it might take four or five days for one person to sucker an acre of tobacco. It's obvious why MH-30 is often cited as one of the most important innovations in tobacco production in living memory.[10] Some growers apply MH-30 just before topping and some just after. As Martin mixed chemicals in the tank of the highboy—including MH-30 and Orthene, an insecticide—he explained that he likes to spray right after he tops because if MH-30 is applied before topping the tops will droop. In addition, safety concerns necessitate not walking through the tobacco for a couple of days after spraying.

When MH-30 was introduced, it, like other chemicals, was applied with a backpack sprayer. Now the suckers are controlled by spraying MH-30 with a highboy, one farmer driving through the fields spraying multiple rows at a time. The highboy—a piece of farm equipment used in many crops, with long legs that enable passage through the rows, above the crop—was also introduced to tobacco farms in the 1960s, but at that time few farmers could afford to purchase one. Those who could afford to invest in a highboy—either by themselves or together with other farmers—often did side work spraying tobacco for those who couldn't. Once sprayed with MH-30, the leaves of the tobacco plant continue to grow, adding weight to the plant. They then begin to turn yellow, signaling that the tobacco is ready to be cut.

Chapter 2

The Harvest through Preparation for Market

In preparation for cutting, *tobacco sticks*—wooden sticks, about four and a half feet long and three-quarters by one inch in diameter, hand-split or manufactured, that will hold the cut tobacco—are dropped in the field. Most often they are loaded onto a platform on the back of a highboy where at least two people stand or sit (depending on the highboy), dropping them into the rows one at a time as the highboy is driven through the field. To ensure the efficiency of cutting, the sticks must be dropped at precisely spaced intervals. According to Roger Perkins, a Franklin County grower, the sticks should be dropped so that they lie in the rows end to end, a few inches apart. Dropping extra sticks results in wasted labor because once a field of tobacco is cut, the leftover sticks must be picked up, and labor has been wasted both dropping them and picking them up.

One day in September, I stood behind Martin while he drove the highboy, and two men dropped sticks into the field. I asked him about the days before highboys, when sticks were carried into the field, one shoulder load at a time. In response, as he often does, he told me a story. One year, he started dropping sticks and had only dropped about two or three rows when a guy came along wanting to cut tobacco for him. At cutting time, growers commonly have to accept labor when it's offered in case they don't get more offers. He told him to go ahead, and the guy started cutting in the rows where Martin had dropped sticks. Then two more guys came along wanting to cut. Martin said to go ahead, and they started cutting. He had started dropping sticks at around one o'clock in the afternoon, and he continued to drop them, enough to keep up with all three cutters, until dark. When they finally quit, he collapsed on the ground and said, "Tell my wife I'll be home as soon as I can get back up." He ended the story with the coda, "Not one of them ran out of sticks." He told me they had cut two and a half acres, which means he carried about thirty-two hundred sticks into the fields on his back at top speed, and "not one of them asked for a stick."

Although mechanical harvesters have been designed and tested since the late 1950s, with the exception of the tobacco on fewer than a handful of large farms where these machines are being tested, burley tobacco continues to be cut entirely by hand. Flue-cured tobacco, which is harvested by grade (*primed*) rather than by cutting the whole stalk, was mechanized in the late 1970s, a decade after the adoption of mechanical topping technologies, as well as bulk curing and loose-leaf marketing, cutting labor costs from 370 to 58 man-hours per acre.[1] Most burley farmers told me they'd like to see cutting mechanized, but all of the attempts so far have resulted in too many damaged and wasted leaves. The mechanical harvesters that are commercially available range from a small machine that is pushed through the field, cutting the tobacco stalk and leaving it for another person to come behind and spear, to a large machine driven through the field that cuts each stalk and drops it on a wagon that must be driven alongside the machine. The stalks are then speared and hung in the barn. These and other mechanical harvesters have not proven to be worth the cost to purchase and run them, particularly since they still require varying amounts of physical labor.

The most mechanically successful burley harvester is the GCH Gold Standard Harvesting System. However, this machine is far out of the economic reach of burley farmers, with an initial investment of one to three million dollars to cover the costs of the machine and the curing structures required for a potential one-hundred- to two-hundred-acre crop of tobacco. This machine cuts the tobacco, spears it, and hangs it in a portable curing structure that is set in the field and covered. These structures hold 448 plants, and therefore fifteen to seventeen structures are needed per acre, at around a thousand dollars each. Mark Roberts, a Henry County farmer who tested the GCH for the second season in 2007, told me that "if it was affordable it'd be worth it. There's no question." It costs him about four hundred dollars per acre to pay a crew to cut, while the machine requires just one person to drive the harvester and another to move

Burley tobacco continues to be cut by hand—hot, backbreaking work. (Photo by author)

the curing structures around. Although the general consensus is that someday burley will be cut by machine, most express the view that it will not be any time soon.

Cutting usually takes place in August, and it is hot, backbreaking work. Each cutter goes into the field with a *tobacco knife* or *tomahawk* (as it is more commonly called) and a *spear.* The tomahawk is an axlike tool, with an eighteen-inch wooden handle and a four-inch metal blade, and the spear is a metal cone, about six and a half inches long, with an extremely sharp point. The cutter picks up a tobacco stick, thrusts it into the ground at an angle, puts his spear on the upward end, grabs a plant with one hand and cuts it off at ground level with the other, and—with the tomahawk still in one hand—grabs hold of the stalk with both hands and spears it about one-third of the way up the stalk, onto the tobacco stick. He does this six times per stick—standing between two rows, he alternates cutting a plant from each of the two rows, cutting what is called a *stick row.* Six stalks per stick has become the accepted norm since wage labor has come to dominate because the workers are paid by the stick, necessitating a consistent number. Prior to this, the number of stalks per stick was often six but might vary with the size and weight of the crop.

In a day, a very good cutter cuts twelve hundred to fifteen hundred sticks, with six stalks per stick, which is about an acre. Stories are told of men who can cut up to two thousand sticks of tobacco in a day—or of men who could in their prime. When the work is done well, the cutter never stops, never stands upright, and never puts down his tomahawk; the sticks do not fall over, and the stalks do not split out. Good cutters are variously described as lean, agile, fast, but steady men. Women rarely cut tobacco, and when they do the exception is noted. Alice Baesler, who raised about three hundred acres of tobacco with her husband in 2007, told me that one of her cutters is a woman and that she can cut one thousand sticks in a day, "and that's good in anybody's book." She went on to say, "Every once in a while you'll find a woman that'll be out there cutting, but most generally, that is a man's job"; "Your good cutters are usually, they've started when they were young, and it's just a thing they learned from their daddy or something and, they've got the technique down." Noel Wise told me that when he was growing up, the ability to cut one thousand sticks made a boy into a man, and in our interview he said, "You grade a tobacco cutter by the amount of sticks he can cut in a day's time. And basically the good strong boys that had experience in cutting tobacco if he could cut a thousand sticks a day he was a real he-man. I don't know how it works today but us old people cut four or five hundred sticks and we give out, but we used to cut a thousand." Garrard County farmers Jonathan Shell (nineteen years old when I met him in March 2007) and his grandfather G. B. Shell described cutting to me this way:

Jonathan: You know, you just don't come out and pick up a tobacco knife and start cutting tobacco. You know, tobacco is such a brittle, plant, you know you—you've gotta get it on a certain position on the stalk, you know, maybe not the exact position, but you've gotta get it in the right position where you don't split the stalk out and it just falls right off the stick. You've gotta, have your stamina up enough to where you can cut a hundred and fifty, sticks in a row.

G. B.: Or two hundred—

Jonathan: Or two hundred or three hundred even.

The Garrard County Tobacco Cutting Contest is the last of such contests in Kentucky. (Photo by author)

Jonathan went on to say, "And the main thing is stamina and form," to which G. B. responded, "It's kinda like basketball or football, the best guys, win." I asked, "Well what makes the best guy?" and G. B. responded "strength," and Jonathan said, "stamina." Later in the season, when Jonathan Shell showed me how to cut tobacco, he told me how much he loved to cut because he feels like he is really working, and at the end of the day he can see what he has accomplished. He pointed to a muscle in his inner thigh and said *that* was the muscle that he felt the first day of cutting season, but it strengthened as the days went by and his body adjusted.

Men often reminisce about cutting tobacco when they were young, racing brothers or friends down the rows. Informal competition in the tobacco field led to formalized cutting contests in which performance was central rather than a secondary function. The audience was no longer limited to other cutters but included possibly hundreds of spectators. In our discussion of the Garrard County Tobacco Cutting Contest, established in 1981, extension agent Mike Carter told me, "We've had very few people enter that are not good cutters. Because, most of them are aware of the level of competition and they don't want to embarrass themselves." Such contests formalized accepted understandings of what it meant to be the best tobacco cutter and therefore to perform one's masculinity for others in a metaperformance. "Best" and "fastest" cutter are not synonymous, and cutting-contest organizers developed very detailed scoring systems that take not only speed but accuracy and neatness into account. According to Carter, prior to the establishment of the Garrard County Tobacco Cutting Contest, conversations about who was the best cutter regularly took place at "the little country stores, and poolrooms, and you know, Farm Bureau meetings."

Men are so associated with cutting—or rather associate themselves with cutting so strongly—that it was not until the 2007 cutting season that I realized that most of the men I had been talking with, interviewing, and sometimes working beside no longer actually cut any tobacco themselves. This is due in part to the aging of farmers. Because cutting tobacco is so strenuous, men cut less tobacco the older they get, and as there are fewer young farmers, it follows that fewer farmers cut their own tobacco. However, burley farmers also cut less of their own crop today because many of them are growing much more tobacco than they once did and therefore spending more time in a management role. This is one changing cultural practice that exemplifies the changing relationship with the crop, as more and more farmers oversee large acreages of tobacco production rather than doing the work of a small patch themselves. As I will discuss, most of the work on tobacco farms is now carried out by Latino men and sometimes women.

Meanwhile, however, farmers' language continues to connect them directly to the task. I was told things like, "I [or we] hope to start cutting next week," or "You're welcome to come back when we're cutting." Although I was well aware that Latino workers were hired during cutting season, I later realized that I had assumed from their self-inclusive language that the farmers were cutting beside the hired men. My assumptions were also based on the fact that many of these same farmers physically participated in other aspects of production, such as setting and topping. Many smaller farmers do still cut tobacco, but it is becoming increasingly less common.

Most growers leave the sticks of tobacco in the field until the leaves wilt, usually for three days.[2] When green, according to Martin, tobacco weighs sixty to eighty pounds per stick.[3] Wilted tobacco is somewhat lighter, and it is less vulnerable to leaf loss and therefore easier to house. Of course, while standing in the field for those three days, tobacco is vulnerable to weather, and I have been told many stories of hailstorms and floods destroying a crop that had just been cut. At the very least, rain on cut tobacco results in *muddy tobacco*—mud-splattered or mud-caked tobacco that, once cured, creates a suffocating dust when the leaves are stripped off; this tobacco may be unsellable. Franklin County farmer Roger Perkins remarked to me one morning, as we watched his crew cutting tobacco, that his grandfather would never have been cutting on this day because there was a slight chance of a thunderstorm that night. According to Roger and others, because tobacco is now often grown in large acreages, it is impossible for many farmers to wait for the ideal weather, and growers just do the best they can. While tobacco wilts in the field, it is also susceptible to being *sunburnt* if it receives too much sun, but I was told by many farmers that three morning dews will take the sunburn out of it.

After the sticks of tobacco have wilted in the field, they are loaded onto wagons and driven to the barn. Carolyn Taylor described a common practice that she and her daughters were responsible for in the 1950s and '60s during tobacco cutting and housing season: gathering dropped leaves. According to Carolyn, "So early in the morning, we would go to the field, the girls and I, and we'd pick up any leaves that we thought were good. And this is when we were not on poundage see this is when we were on acreage and you wanted to get every bit of tobacco off of the field that you could." She went on to explain that they put them in a "big sheet" and "laid them in that and then we just four-cornered it and brought it in the barn and then we'd come to the house, and start cooking." According to her daughter Diana Taylor, they would then string the leaves on wire and hang them in the barn to cure, and Carolyn and her daughters would receive the money for those leaves when they sold. This is the only form of cash compensation that Diana remembers receiving when she was growing up, although she and her sister participated in other aspects of tobacco production and performed other farm and household chores. Others described picking up the stray leaves and stringing them onto sticks. Once the quota system changed to a poundage system, individual leaves were of less importance, and this practice gradually waned.

Although they may not physically cut tobacco themselves, many farmers drive the tractors that pull the wagons onto which the tobacco is loaded. At this point the *housing* begins. Housing tobacco refers to hanging the tobacco, still on the stick, in the barn, where it will hang until it is cured and ready to strip. Presumably the term *housing* comes from the former term for tobacco barns, *tobacco houses.*[4] Different types of tobacco, because they require different curing methods, are housed in distinctly different barns.[5] Dark fire-cured tobacco, for instance, is cured in barns that are much smaller and taller, so that the smoke from smoldering fires built below the hanging tobacco from sawdust and slabs of wood can cure the leaves. Flue-cured tobacco was once cured in small barns with indirect heat but is now cured in rectangular metal barns called *bulk barns.* Burley tobacco barn sizes vary, as do descriptions of barn sizes. Barns might be described in terms of their number of *tiers, rails, bents,* or even *posts*—most often by rails or bents, rarely by such conventional measurements as feet or inches. The posts and tiers section off the barn into bents, which are usually twelve- or fourteen-foot sections of the barn, separated by support posts stretching from ground to roof. The horizontal beams, which stretch from post to post across the barn, support the rails where the tobacco will be hung; each layer of rails is referred to as a tier. Tobacco barns usually have three tiers, vertically spaced four to five feet apart, and may have as many as five; there is often an additional tier above the central corridor of the barn below the peaked roof. Burley tobacco barns are usually lined with vertical side vents in order to allow for the control of ventilation and moisture levels inside the barn as the tobacco cures, although occasionally barns with horizontal vents along the top or bottom can be seen dotting the landscape. The central corridor, with doors on each end, is large enough for a tractor pulling a wagonload of tobacco to drive through. While conventional burley barns can hold two to five acres of tobacco, the largest barns can hold ten to twelve.

When housing begins, a load of tobacco is driven into the barn. Several workers climb up into the tiers, and the tobacco is handed up to them. A crew may include one worker per tier on each side of the wagon, which means that in a three-tier building there may be six workers up in the barn. A smaller crew might work on one side of the barn at a time, cutting the number of workers in half. Others usually stand on the wagon and hand sticks of tobacco either directly up to those in the tiers or to workers on the ground, who then hand

A typical burley tobacco barn is lined with vertical side vents that can be opened and closed in order to regulate moisture during curing season. (Photo by author)

it up into the barn. If women are involved in housing, they are most likely to hand tobacco up to men in the tiers; they rarely climb up in the barn. When I watched tobacco being housed at the Gallagher farm in Fleming County, the crew housed about six double wagonloads in three hours. This crew of ten workers included four up on the rails of the three-tier barn and six handing the tobacco up from the wagon: three per side, with two on the wagon, one on the ground at the bottom tier on each side, and one between the wagon and bottom tier on each side. This was an abnormal crew, made up mostly of local workers, all white, and including three women.

Latino workers, primarily from Mexico and male, now do the vast majority of cutting and housing on most Kentucky farms, much as they now do much of the harvest work on farms across the country. Their work accounts for at least 75 percent of total labor hours.[6] The 1980s was a transitional period in which more and more farming men and women were taking off-farm or *public jobs* in order to supplement farm income. At the same time, it was becoming more difficult to find local labor for hire that was willing and able to work

Workers housing tobacco in the top tiers of a burley barn. (Photo by author)

seasonally and dependably on the farm. These and other factors combined to create a severe labor shortage. One response to this shortage was a renewed push for mechanization, and the second was the recruitment of migrant laborers, primarily from Mexico.[7] Beginning at the end of the 1980s and the early 1990s, Latino workers became the primary labor source on Kentucky farms and came to be viewed as their saving grace.

Most Latino farm workers are undocumented migrant workers, although as farm sizes grow, tobacco farmers are increasingly participating in the federal guest worker program known as H-2A, which regulates the legal hiring of noncitizen farm laborers. Not all Latino workers can be classified as migrant, as some have settled into Kentucky communities. Those who do stay often find employment that pays more and is more consistent than farm work. Farmers have commented that this is because farm workers become "Americanized" and no longer want to do physical labor, although others tell me that many find other jobs that pay better, such as construction work. It is impossible to generalize about the relationship between Kentucky farmers and Latino workers, upon whom they depend.

Farmers frequently express their feeling that Latino labor saved them. They would not be farming today, many told me, were it not for Latino workers. According to many farmers, Latino workers are willing to work hard for long hours, unlike those local workers who are available to work seasonally—workers who cannot hold full-time, year-round jobs for a variety of reasons. Alice Baesler described her experience trying to hire local workers:

> And the one thing that kills me . . . is when people say "they're taking our jobs." Well, and then they also say, "Well . . . people wanna hire them because they're paying such a low salary." Now wait a minute. I'm paying the same if not more, for this H-2A laborer than I am for the American labor. . . . We last year, put an ad in the paper, and we had thirty-seven responses, and we told them we would call them when we were gonna start. We called them. We had five of the thirty-seven show up. The next day we had one.

According to Alice, the required hourly wage that year was over nine dollars. I asked about the other expenses of the H-2A program, and she responded: "When all of those things are added in—the housing, the transportation, what you pay to bring them up here. And, going to the grocery every weekend and all that. It adds up to $14.50 an hour. That you are paying. Which is very expensive farm labor. This is why it's so difficult for some of the really small farmers, to do that."

Many farmers who rely on Latino labor see a core group of the same workers return every year, and many have formed close relationships with these employees, some even taking trips to meet their families and see their homes in Mexico. At the same time, however, some farmers claim that the quality of their tobacco has suffered because migrant workers do not handle it with care, and on occasion I heard overtly racist remarks about the ability of these men and women to perform as expected, including characterizations of them as children who can't be trusted to follow directions.[8]

In the 2006 crop year—one with particularly high yields—farmers who depended on migrant labor faced a severe labor shortage that many attribute to the political focus on undocumented workers.[9] The Immigration and Naturalization Service had long turned a blind eye to undocumented farm workers, but 2006 brought multiple farm raids and stories of undocumented workers being picked up by the authorities while traveling to Kentucky farms. It is thought that fear also kept some workers from coming into the United States that year.

There are numerous contract arrangements between farmers and workers. Some contract directly with individuals, some contract with a crew through an agent (also usually a Latino man) who manages multiple crews, and some contract with and house the same crew for the entire cutting season (or longer). According to the scenario described to me by both farmers and extension agents, the politicization of immigration issues that came to a head in 2006 resulted in a shortage of workers in all but the last category of contract relationships. Because there were fewer workers, crews and individuals who contracted by the job rather than the season committed themselves beyond what they were able to handle. I heard stories about crews who cut tobacco and then did not come back for well beyond the desired three days to house it, increasing the chances of muddy or sunburnt tobacco. Farmers tell stories of some farmers watching their crops rot in the field in 2006 (as well as in more recent years) because of a lack of workers to cut it or to house it once it had been cut.

The workers whom Martin hired in 2006 came about fifteen days later than he wanted. They cut half a day, and then six inches of rain fell, resulting in very muddy tobacco in that particular field. He was able to save some of it, but much of it had to be discarded, including a disproportionate number of the most valuable leaves, the *tips,* or uppermost leaves. Martin said that it took twice as long to strip that tobacco (therefore doubling his labor costs) and described cleaning the dusty stripping room out two or three times a day. In his typical glass-half-full manner, he laughed as he said that he was just glad that they had only cut for half a day rather than a whole day.

Clarence Gallagher also had a hard time in 2006; he ended up himself housing the tobacco that a crew had cut, because they didn't make it back in what he saw as an appropriate length of time. As a result, he was determined to get together a crew of local workers in 2007, a group that would do all the cutting and housing together from start to finish so that they would be able to work at the same pace. He told me that someone joined them one day who was not working at their pace and couldn't keep up (he was "green"), but other than that this crew stayed together through the cutting and housing of Clarence's twenty-two acres. Clarence is the only grower

I met who had a crew of local, all-white labor in 2007. The crew, whose roles I mentioned above, included his college-age son and a couple of his friends (including one young woman), two women from an Amish community who have worked with him throughout the tobacco season for a number of years, two men with whom he swaps work, a couple of other local men in their forties and fifties, and Clarence himself.

Housing would once have been done by local boys—for instance, a farmer's son(s) and other high school boys. The shortage of local help that began in the 1980s and peaked in the 1990s is commonly blamed on new employment opportunities for teens, such as fast-food restaurants. I was asked repeatedly, "Why would kids want to work in tobacco in the hot summer if they can flip burgers?" There are few fast-food restaurants or other such job opportunities in the community where Clarence farms, and I suspect that this played an important role in his ability to get this crew together.

There is a general feeling that the labor shortage is one of the greatest problems growers face. I've been told by some farmers that lack of labor may be what finally forces them out of tobacco, and it is generally commented that if there were a satisfactory and dependable solution to the current labor shortage, there would be no limit to Kentucky tobacco production (except, of course, the limits imposed by the market). Not unlike during the labor shortage of the 1980s, mechanization is once again being looked at as a possible solution, leading to the belief held by some that mechanization will eventually take over, and small growers will be completely forced out. There has been much discussion about whether the tobacco companies (mainly Philip Morris) are trying to push out smaller growers so that they have fewer growers to deal with or whether they recognize that they need the higher-quality tobacco that small growers continue to produce over larger growers who, most agree, have traded quantity for quality.

Housing is dangerous work, and many tell stories of people (including Martin) falling out of barns during housing. The sticks should be spaced about six inches apart so that air can circulate around the leaves as they hang upside-down from the sticks, and—particularly if the tobacco is tall—the stalk ends should be carefully alternated with the tops of the tier below so that they do not overlap. Martin prefers that his tobacco be hung eighteen sticks to the rail, and he says that after the tobacco has hung in the barn a few days, you should be able to see the underside of the roof when standing in the barn. Proper spacing prevents *house-burn,* which results in rotten stems and leaf loss; according to Martin, you can smell it as it starts. Occasionally, a stalk of tobacco falls when the stalk splits out because it was not speared in its center (a sign of bad cutting that Jonathan mentioned in the interview exchange quoted above). The grower or other supervisor usually picks these up and respears them onto new sticks. These are the kinds of practices that some growers told me are getting lost as labor has moved from the family to paid workers.

CURING AND STRIPPING

Burley tobacco is air cured, in contrast to heat-cured tobaccos such as dark fire-cured (produced primarily in western Kentucky and Tennessee) and flue-cured (produced primarily in North and South Carolina, Virginia, Florida, and Georgia). The weather must cooperate in order for burley to properly cure. If the weather is dry, it will cure too quickly and will be in danger of getting *piebald,* or too dry and yellowed; more time in the barn in humid weather can remove signs of piebald. If it gets too cold before it is fully cured, it will stay partially *green,* never fully curing. If it freezes, it will get *fatty stems*—stems that do not cure up and literally feel like animal fat. It is hard—often impossible—to reverse the damage done to green or frozen tobacco.

Traditional burley tobacco barns are distinct because of the vertical shutters that line their sides. In the past, these shutters were methodically opened and closed in order to regulate the curing and ensure that the tobacco came *in* and *out of case* or *order* daily, which is what must happen if tobacco is to cure and not simply dry. When tobacco is *in case* or *in order* (*case* is used more frequently in central Kentucky, so I will use it here), it is moist and pliable; when it is *out of case,* it is brittle and crumbles at the touch. Shutters are not always opened and closed anymore, a traditional practice that farmers often point out as lost, some attributing the change to laziness, others to size, as large growers may have twenty or thirty or even more barns full of tobacco, spread out over miles of farmland. Still others attribute the change to shifts in the style of tobaccos that manufacturers want. The recognition of the waning of this practice became evident during the 2007 curing season, when the

University of Kentucky College of Agriculture reminded farmers of the importance of using "management techniques" to hold in what little moisture there was. A September 2007 press release included this advice: "The unrelenting heat that has held a tight grip on Kentucky gradually will submit to a series of disturbances that will move through the region during the next several days."[10] The release went on to say:

> The potential for significant rainfall is not as bright, but chances of showers are in the forecast for the next several days.
>
> As a result, tobacco growers should try to open vent doors at night or during periods of high humidity, and keep them closed in dry periods during the day.
>
> "The idea is to allow the moisture of evenings and rainy days to migrate into the curing structure and bring the tobacco into 'case' or 'order' which means leaves being in a pliable or non-shattering condition," said Gary Palmer, UK Extension Tobacco Specialist.

Numerous alternative curing structures have appeared in recent years, many through university research, some adopted more widely than others. The uncertainty of tobacco's future has kept many farmers from building expensive new tobacco barns, and many have instead begun to use field-curing structures—scaffolds made of wood or a combination of wood and wire built at the edge of the field on which one tier of tobacco is hung and then covered with plastic.[11] Such structures came into use beginning in the mid-1990s and are more economical in the short term. According to agricultural engineer George Duncan, they can be constructed for one-tenth of the cost of a barn. Many farmers have told me that it can be difficult to keep the plastic on, resulting in the curing process being interrupted as the tobacco is exposed to the weather. Others have built barns designed to be more economical, such as new models that are only partially enclosed. Although much has been made in the local and national media about "what to do with all those empty tobacco barns," and many barns are in fact suffering from ill repair, farmers actually face a shortage of barn space. [12] Barns are now often not in the places where they are needed—with post-buyout shifts in the location and size of tobacco fields—and owners of barns in disuse are increasingly unwilling to rent them out either because of safety concerns or because they just don't want people coming onto their farms. In addition, farmers are reluctant to spend the money to maintain barns in the current climate of ongoing uncertainty. Mark Roberts of Henry County told me:

> It's kinda funny, here we are—back then, that was what? Ten or fifteen years ago or longer we were saying "Well I don't know where tobacco's going, I'm not sure if it's gonna be here, and I'm not sure if we're gonna get paid enough. We can't afford to build barns—we gotta look at something else." Well here we are right now, this many years later saying the same exact thing. "Is it gonna—is it always gonna be here, do they always wanna buy our tobacco? And are they gonna be willing to pay for it?" So that hasn't changed.

Curing usually takes place between August and November, with tobacco stripping starting between sometime in October and Thanksgiving, depending on the weather, when the crop was set and topped, and the amount of tobacco that a farmer raises. Some large growers start stripping tobacco in September and in some cases may be able to *double-crop* their barns—allow a crop to cure, take it down and strip it, and then house a later crop in the same barn. The goal has traditionally been to finish stripping by Christmas, both in order to get paid for the crop in time to pay end-of-year bills and to buy Christmas gifts and because farmers have noted that after the Christmas break prices often drop as tobacco companies begin to see their demand fulfilled. According to Anderson County farmer Phil Sharp, "Well the quicker you could get it to market the—it's just like somebody going to the table, after they get so much they don't get as hungry as they do, when they about got their bellies full, and maybe if you was on the tail end of the market you wouldn't do quite as good if you was if you was on the first end of the market." Many farmers also have winter calves being born in January and February and generally have to feed cattle up to twice a day, so attention needs to be turned there. The shift from auctions to direct contracting has meant that farmers now have much less control over when they can bring their crop to market, as I will discuss.

Ultimately, the weather dictates when tobacco can be stripped, and I've been told of years that farmers didn't get their tobacco stripping finished until as late as April, as they waited on the weather. Patsy Perkins of Franklin County told me about a year that they finished stripping tobacco just in time for her to sew Easter dresses for her daughters. Martin didn't begin stripping his 2007 crop until December, in part because it was late, as described above, but also because he had to wait until the weather was right to take it down out of the barn, or *book it down*. The dry early fall that most growers experienced became a wet late fall and winter for Martin—circumstances from which he benefited in the end, as his tobacco cured well. It became the dark brown-red color that the companies want to buy rather than the tan, *yellow*, or *bright* tobacco that so many growers ended up with. Robert Taylor told me, "Ideally, you want it to be the color of an old penny. Not a new penny but an old penny—if it was that color you were in business." Martin told me this about his 2007 crop:

> Martin: I had one thing in my advantage, I did have a red crop, whereas, if it'd been earlier, it might have piebald. "Piebald" is where you—it doesn't cure up it dries up.
> Author: Which is what most people have right?
> Martin: Yeah that's what most people have. As a matter of fact, I took a load of tobacco to the warehouse last week and, the buyer said "Well, we got a load of *burley* tobacco today!" [*laugh*]

Having good-quality tobacco is always a source of pride, and in the 2007 crop year this was even more the case because it was so rare. This meant even more to Martin because of the way it had looked earlier in the season. The day I accompanied him as he was dropping sticks, he pointed out a particularly tall patch of tobacco. I asked if in better years it all looked like that. "Oh yes," he said, adding, in reference to most of his crop, "This is the commonest tobacco crop I've had in years."

Martin wasn't the only one commenting on his crop. One day in December, I was visiting with a farmer on Martin's side of the county, and we were joined by another farmer. They talked about the poor quality of most growers' tobacco that year, and then the second farmer (who may or may not have known that I knew Martin) said, "I'll tell you who is gonna have some good tobacco is Martin

Henson." He went on to describe Martin's tobacco as being late and therefore small and dark—just what they wanted that year—and noted that Martin had stripped some the previous week but had not gotten any more down since. Martin won't touch his tobacco if it's not just right, he said.

Stripping tobacco is the process of removing the cured leaves of tobacco from the stalk and separating the leaves into grades before packaging the tobacco for sale. Stripping begins with *bulking* or *booking*[13] the cured tobacco, which means climbing back up into the tiers of the barn and handing the sticks of tobacco down and placing them in a *bulk* or pile of tobacco. Sometimes this is simply called *getting the tobacco down* or *putting it down.* The tobacco is taken off of the stick and stacked on a wagon, sometimes in tied bundles and sometimes loose. The empty sticks are stacked in the barn—also sometimes in bundles, sometimes loose—for use the next year.

As described above, the tobacco must be in case when it is booked down so that it can be handled without being damaged, but it cannot be *too casey* or in *high case* (too wet) because then the risk is that it will rot after it has been stripped and baled. According to Jerry Bond, "You have to, be knowledgeable enough." He went on:

> When you just go out and feel of it, if it's too dry, or too wet, or just right. And some people can do that and some can't. And the ones that can't suffer when they get to the market because if it's too dry you're giving up poundage. Because wet tobacco's gonna weigh more—it's absorbed moisture. If it's too wet, they won't buy it. So it's got to be just right. And there are a few people—most everybody around here can—but there are some that don't have a clue when they go out in the barn, and grab the tobacco and know if it's ready to be stripped. It's an art. It's an art.

Tatham's 1800 description suggests that judging when to book down has long been considered a specialized skill. He describes the term "in Case" as "a technical term made use of by the planters to signify a specific condition of the plants, which can only be judged of safely by long experience."[14]

Often tobacco is at its best in the middle of a damp night, and farmers such as Jerry Bond told me stories of their fathers waking them up in the middle of the night to book the tobacco down; sometimes they also began to strip it then, and sometimes they waited until morning. This is yet another aspect of tobacco production that has changed with the changing labor situation. For farmers who depend on paid labor rather than family or community-based labor, it is most often not possible to get the tobacco down at the optimum time because it is not practical to expect paid laborers to come to work on short notice in the middle of the night. Some growers, particularly large growers, have come to depend on methods of bringing tobacco into case so that it can be handled at any time, such as spraying it with water (a practice other growers describe with disdain). Loads of tobacco are often covered with plastic once it is down in order to hold in the moisture until it is brought into the stripping room.

I arrived at Martin's one cold, damp Saturday morning in December. Martin had gotten his crew started stripping tobacco around 7 a.m., and he was in the process of feeding cattle. I rode in the truck with him, staying put each time he got out to fill the feeders with buckets of feed, with Buster asleep on my lap. The deep winter mud had frozen, making it difficult for Martin to walk; he'd had ankle surgery in October and was still wearing a cast protected by a large rubber boot. A farmer whom he often swaps work with, Elic, had been helping him around the farm since his surgery. After he finished feeding, he and another man who'd turned up started to work on a highboy that wouldn't start and that Martin needed to move into his equipment shed, so I rode over with Elic to feed his cattle on a nearby farm. By the time we returned, they had the highboy running and were moving on to a tractor that was giving Martin problems, so Elic drove the highboy over to where it needed to go, and I followed in his truck to bring him back. Then Martin headed out on the tractor to feed hay to the cattle, and Elic got me started on tips in the stripping room.

Martin's stripping room is set up much like others in which I stripped tobacco or visited and like those described to me many times from memory. Traditionally, a stripping room is a small narrow rectangular room, about twelve feet wide by twenty feet long, attached to a tobacco barn, in which a farmer's tobacco is stripped from the stalk and readied for sale. A great deal of nostalgia pervades descriptions of stripping rooms of the past, whether given to me in interviews or in brief conversations about my research. Families spent the fall and often much of the winter together in the stripping room, and these times are described as multigenerational, full of stories, warm from the stove (wood, coke, or gas), and pervaded by the pleasant smell (as well as the less pleasant dust) of cured tobacco. In most cases the entire family was involved—all ages, all genders. Dinner and supper were sometimes cooked or reheated on the stove, the smallest children played or slept on a corner of the stripping bench, a dog usually lay under the stripping bench, neighbors stopped by, and the radio (later sometimes a TV) played in the background. I commented to Phyllis and Phil Sharp during our interview that people often talked about stripping rooms as important places of family memories, and Phyllis responded, "Oh yeah it was like a gathering place in the wintertime with the neighbors, you know. If they weren't stripping or they just wanted to come and hang out, they would come and, you would have like a big pot of beans and cornbread or a big pot of chili, a big pot of soup on the old stove in the stripping room. And the radio going and it—yeah, it was hard work but a lot of good memories and a lot of quality time with your family."

Most conventional stripping rooms have the same basic floor plan: a wooden bench about three feet wide and waist high runs along the length of one wall. There is usually a door leading into the stripping room from the barn, where wagonloads of tobacco wait and where tobacco stalks and finished bales will be brought. The wall parallel to the bench usually houses the stove and, since the early 1980s, a set of baling boxes and an air compressor. Some stripping rooms, including Martin's, also now include sets of wooden boxes in the middle of the room where tobacco is put, by grade, between the time it is stripped and when it is baled. Instead of baling boxes, stripping rooms of the past featured a wall-mounted tobacco press, where sticks of tobacco tied in hands were flattened in preparation for sale. The replacement of the wall press with a set of baling boxes represents the most often cited change in burley tobacco production: the move from hands to bales.

Armfuls of cured tobacco are brought from the wagon into the stripping room and laid on the bench. The leaves are stripped off into grades, which are determined based on a combination of stalk position, size, color, and texture. At one time, a single stalk of burley tobacco was stripped into as many as seven or eight grades; three is now the norm, although the companies would prefer four. The names of grades of tobacco differ somewhat by family, community, and region, and vernacular names differ from those once used by government graders, who had a far more complex and numerous range of grades than those used on the farm (as many as 109 at one time) and those used by tobacco companies (each of which has its own system). For instance, Charlie Long of Hart County remembers these seven grades from his younger days: *buzzard trash, good trash, lugs* or *flyings, tan leaf, red leaf* (sometimes *short* and *long*), *short tips,* and *long tips.* Martin remembers *trash, lug, long lug, short red, long red, tip,* and *short tip.* Today's grades—which have dominated for the last couple of decades—are fairly consistently known as *trash* or *flyings, lugs* or *cutters* (it is this middle grade that Philip Morris would like to see divided into two, resulting in four grades), and *tips.* Some farmers also have a *crop throw* grade in which they bale together their lowest-quality leaves of all grades—green leaves, piebald leaves, leaves swept from the stripping-room floor—in order to improve the overall quality of their other bales in hopes of a higher price. They'll get something for their bales of crop throw, however small. It seems to be becoming more common to refer to the grades by number, perhaps because of the language barrier between many farmers and the Latino men (and sometimes women) who strip tobacco for them. I often heard farmers refer to the three grades as *ones, twos,* and *threes* when talking with Latino workers.

There were periods in the 1980s and '90s when many farmers stripped their tobacco into only one grade, and opinions differ on the reasons for this and the outcomes. I've been given two explanations. Mostly I've been told that the demand for tobacco at the time was higher than the supply, and so the tobacco companies would take whatever farmers had to offer, the implication being that farmers had the upper hand, however briefly. But I have also been told that buying tobacco in one or two grades was strategic on the part of the companies. In one version of this explanation R. J. Reynolds was responsible for tobacco being graded into two grades because they could use it all, and in another Philip Morris actively encouraged farmers to strip their tobacco into one grade in order to gain control of the market. While other buyers, particularly smaller leaf buyers representing European interests, wanted particular grades of tobacco in specific quantities and therefore were not interested in buying mixed-grade tobacco, Philip Morris could use it all in their blends. This version reverses the power relationship between the farmers and the companies, while also serving as an origin narrative for how Philip Morris came to control the burley market. Regardless of the reason, some argue that this period marked a low point in the quality of Kentucky burley, and some farmers continued to strip their tobacco in multiple grades even though the market at the time did not pay them for it.

I asked Martin about the period in which tobacco was stripped into one grade, with some surprise that he'd never mentioned it to me in a year of work and talk, and he told me, "Yeah. I never did do that. But yeah, back in the, in the mid-'80s, you could strip it in three grades and get the same amount of money that you could in, two or one. That was, I don't know three or four years. And a lot of people—of course it cost money to, strip it." It cost at least two or three cents per pound more, and "you take eight thousand stalks to the acre, and, pick up eight thousand stalks an extra time, or two, it takes—you know, it adds up." I asked him, "So why didn't you ever do that?" and he simply said, "I always thought there were two or three grades on tobacco."

Burley tobacco is stripped beginning with the bottom leaves (*trash* or *flyings*), followed by the *cutters* or *lugs,* and finally the *tips.* Most stripping rooms are organized so that each person strips one grade and then passes the stalk down the bench to another person who will strip the next grade, but some are organized so that each person strips all the grades off of each stalk; opinions differ on which approach is more efficient. Learning the difference between the grades is obviously the first skill that one must acquire when learning to strip tobacco, and beginners are started on tips since it is not possible to make a mistake when that is all that is left on the stalk. I repeatedly heard versions of the following description, this one given to me by Charlene Long in September 2005: "When you started out, stripping with the family, usually—when you were little you got the tips. You got the end of it because you couldn't mess the tips up. Well when you got up oh around ten or eleven or maybe twelve when you were getting up, you might venture down to the red leaf, or they might put you on the end with the trash. But those would be the next, those would be your next step. But usually your best tobacco stripper was the one that stripped what we call the *lugs,* or the *flying.*" The bottom or *trash* grade is called such because, as I was often told, it looks "trashy."

The name also reflects the changing values of the various grades according to the needs of tobacco companies in various periods. At one time, the "trash" was of no value; later it was the most valuable. It is now once again of least value.[15]

The *trash* consists of the leaves closest to the ground, and they are often crumbly, light in color, thinner than the others. As you move up the stalk, the leaves thicken and become darker until finally you get to the tips, which are smaller and should be a dark brown-red—not "bright" as so much tobacco was at the sale of the 2007 crop. I was usually started on tips in the handful of stripping rooms that I worked briefly in, and I was happy with this, afraid that I would get the grading wrong and have some negative impact, however small, on the quality of that farmer's crop. But the more tobacco I stripped, the more I could see and feel the grades; I could even begin to see how tobacco could be graded into more than three grades. Once all leaves are stripped from the stalks the stalks are stacked on the bench or on a wooden stick rack. Periodically, the stalks are carried out to a wagon and later scattered in the field, where they will degrade and be disked into the ground in the spring, returning nitrogen to the soil. Some growers chop the stalks before scattering them on the field.

There have been numerous attempts to mechanize the tobacco-stripping process, including "knot hole devices" that yank the stalk through a hole and strip off the leaves and "stripping wheels" that rotate horizontally: cups are attached to the stripping wheel to hold the stalks, and workers strip the leaves off as the wheel rotates. Although many of these devices have been tried, thus far none have caught on widely, and most have been abandoned after a short time of use. Many farmers tried mechanical stripping machines during the late 1980s, when there was a general push for mechanization because of a decline in available labor. For instance, the Perkins family of Franklin County bought a stripping machine but stopped using it after just two or three years because it was so noisy that they could no longer talk to one another while they worked, and it was hard to keep up the pace required in order to get the optimum benefits of using it. Tobacco farm families have long had some paid labor in the stripping room—at midcentury, for instance; I am told that this was a job that widows often did. It is becoming increasingly common to pay Latino workers to strip most if not all of the tobacco, as farmers without family help are busy managing other aspects of their farm operations. But even today, retired farmers or farm workers, mostly but not all white men, are paid to help strip tobacco on many farms.

Martin tried to spend at least part of each day at tobacco-stripping time in the stripping room with his crew, and it was obvious that this led to a comfortable and even playful atmosphere. The first day that I stripped tobacco at Martin's, I stood beside him after he finished feeding and joined us in the stripping room, as his crew (two Latino men and an older white man, whom I had met the first time I met Martin) joked with each other, with him, and eventually with me. Spanish-language radio played in the background. As we worked, he told me things that I'd heard from him and other farmers before, for instance, that when he was young they stripped into seven grades and that the trash used to bring the most money, but now it's the tips that the tobacco companies want.

There is a range of explanations about why the companies have valued the grades differently at different periods. For instance, I was told that the tips contain the most nicotine and that once manufacturers of cigarettes were prohibited from adding extra nicotine to their products, the tips became more valuable. In addition, different companies have always bought different styles of tobacco in order to suit the products they manufacture. Generally speaking, however, mystery shrouds the decisions and actions of the major cigarette companies. A January 1994 comment by the commissioner of agriculture makes this clear: "Ever wonder why tobacco farmers seem to be in a continual state of confusion and frustration? Maybe it's because the tobacco companies want it that way."[16] Although he was talking about the fluctuating demand for tobacco at the time, his remark applies more generally to the relationship between farmers and tobacco companies. By keeping farmers guessing over the years about desired grades and desired characteristics within these grades, the tobacco companies have consistently been able to name the price with little or no explanation; this was the case with the tobacco program in place and is even more the case under contracting.

At one point on this first day of stripping at Martin's, he handed me a *hand* of tobacco—he'd tied it so quickly, I hadn't even noticed he was tying it—and said, "This is the way we used to do it. You can take that with you." Leaves of cured tobacco were long tied into what is referred to as a *hand* in preparation for sale.[17] A hand of tobacco is formed as the leaves are stripped from the stem in a particular grade.

The stems are held tightly in one hand, leaves pointed toward the floor. When a handful has been stripped, a leaf of the same grade, called a *tie leaf,* is wrapped around the stems multiple times and then woven through the stems, holding the hand together. Simple as this may sound, tying a "pretty" hand of tobacco—and "pretty" is frequently used to describe how a hand should look—is far from easy. It was a skill that farmers took great pride in because it was understood to be a skill that not everyone had. The stems were to be perfectly even. The sizes of hands varied (depending on the human hand holding it), but smaller seems to have been better at one time—I've heard many descriptions of grandfathers who took pride in the "tiny little hands" of tobacco that they tied. The tie leaf was often a specially chosen leaf of just the right color and texture, and when found, it was often held under the arm until the time came to use it. The tie itself took a particular skill in order to produce an even band around the butt end of the hand and to ensure that the stem end of the tie leaf did not protrude. Often finished hands of tobacco were *skirted*—the best-looking leaves in the grade would be slipped under the tie leaf of the finished hand, forming a skirt around the entire hand; these hands would later be placed on top of a pile of tobacco at sale time.

As each hand was tied, it was hung on a tobacco stick protruding horizontally from a hole in the bench until the stick was full, which generally meant twelve to fourteen hands. Each full stick of tobacco was placed in the press that hung on the wall and flattened; it was then carried out to the barn, where it was booked, by grade, until it was time to take it to the warehouse. When the tobacco was brought to market, the hands were removed from the stick and carefully packed onto tobacco baskets, which are large shallow square baskets, roughly three feet square and four inches deep. It was fanned out in a circular pattern with the tied ends to the outside, into *piles*—still segregated by grade—that weighed between four hundred and seven hundred pounds and were four or five feet tall. [18] Farmers were very particular about the packing of their baskets, both because of the economic advantage that they saw in high-quality tobacco and because of the pride they took in how their tobacco looked. They often waited extra time—and they might have already waited all day or more to unload—to get the warehouse employee they preferred. The piles were moved around with small handcarts called *duckbills,* lined up in rows, and graded by government graders, and each was

KENTUCKY AGRICULTURAL NEWS, February, 1975—7

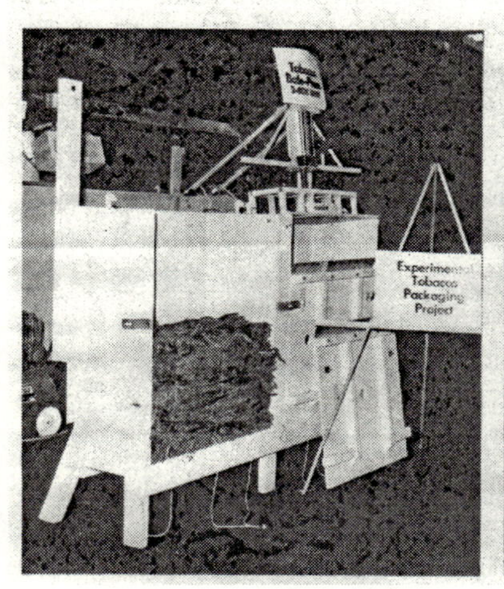

NEW TOBACCO PACKAGING PROJECT DEMONSTRATED

Engineers at the University of Kentucky's College of Agriculture recently demonstrated an experimental tobacco packaging machine (at left) designed to cut down on the amount of labor needed in the stripping and handling of cured Burley tobacco. The leaves are pulled from the stalk and placed in the machine with the tips overlapping. The leaves are then compressed into a bale by the cylinder at the top, which is powered by a conventional air compressor. The bale is tied with three strings to complete the package, which weighs approximately 35-50 pounds. In the photo below, Joe Smiley (left), UK tobacco specialist, George Duncan (center), UK Ag. engineer and designer of the machine, and Harold Harberson, tobacco specialist with the Kentucky Department of Agriculture, inspect one of the bales as it would appear on the market floor.

The design of a machine to bale burley tobacco began in the 1970s, led by George Duncan and Joe Smiley of the University of Kentucky. By 1983, 80 percent of burley farmers packaged their tobacco in bales, and soon after, tobacco could no longer be sold at auction in the traditional package called a *hand*. (*Kentucky Agricultural News*, February 1975, Courtesy of the Kentucky Department for Libraries and Archives)

then auctioned separately. Once auctioned, the hands were packed into hogsheads for shipment.

This all began to change in the 1970s, as specialists at the University of Kentucky, led by agricultural engineer George Duncan and tobacco specialist Joe Smiley, began designing and building equipment for baling burley tobacco. After more than five years of research and experimentation, the move from tying burley tobacco into hands to packaging it into bales happened rapidly—effectively it took just three years, from 1980 to 1983—as compared to the spread of other agricultural innovations. According to George Duncan, the "whole state of Kentucky and surrounding states converted over, in about a three-year period. And that was somewhat phenomenal, at that time to change a hundred thousand growers, in a period of three years." The story of the move from hands to bales is a political story as much as it is a technological one, and one with as many sides to it as there were growers, warehousemen, tobacco buyers, and extension agents—making it much too large a story to do justice to here.

But I will describe some of the parts of the picture as I have been able to put it together.

The University of Kentucky began testing tobacco baling in 1974 with thirteen Bourbon County farmers. In the late 1970s, larger trials involving more farmers were held to compare the labor required to package tobacco in hands, in bales, and in *sheets,* which meant packaging it in large sheets of burlap, as had become standard in flue-cured regions. Whether to package burley in bales or in sheets was a major question for a time, and some farmers did begin to use sheets briefly. In 1980, growers could choose to bale 25 percent of their crop, and bales constituted 12 percent of the burley sold that year. In 1981, the unlimited sale of baled burley tobacco became possible through changes made to the federal tobacco program, and by 1983, 80 percent of burley tobacco was sold in bales.[19] Soon after, buyers would no longer accept tobacco that was tied in hands.

Martin's story of switching to bales is reminiscent of his narrative about moving from plant beds to float beds, and it is quite typical of the stories I heard about the move: he had a baling box made by the students in the Future Farmers of America chapter at the local high school, and he had one crew using it in one stripping room and another crew tying hands in a different stripping room. The crew that was tying hands ran out of tobacco and came over to work in the stripping room where the crew was baling, and "that was it," no one would tie hands after that.

And yet farmer resistance is a common narrative thread characterizing the move from hands to bales. A 1981 letter to the editor expresses this perspective clearly: "This is the worst display of tobacco for market I have seen in my 38 years of growing burley. Packaging tobacco in bales and sheets takes all the pride out of growing a quality crop of tobacco. I suppose this is the way of the world, and what some people call progress. To me, it's taken all the pleasure out of growing and sending a top quality crop to market."[20] The majority of the farmers I have talked to, however, told me that not only were they not resistant, but they were eager to move to bales because of the labor they would save (from 40 to 50 percent), suggesting either that the perspective in the letter to the editor represented a minority of farmers or that farmer acceptance of baling in the intervening years has overshadowed any reluctance at the beginning. County agent Keenan Bishop suggested the latter explanation: "You know a lot of them cite that as a positive change but at the time, you know looking back it saved a lot of time. It made things easier. But when you're used to doing hands and, your great-grandfather tied it in hands and now they want you to do it in a bale, there's a thousand reasons why that won't work."

As I discuss in a later chapter, many of the men I have talked to told me that their fathers expressed doubts about baling; if their fathers did not actually live to see tobacco in bales, these sons believe they would have been appalled. Concerns about baling included fears that the bales would retain too much moisture and would rot, and some worried about "nesting"—when nontobacco matter gets into the tobacco, from unintentional garbage such as candy wrappers to items put there intentionally in order to increase the weight of a bale. Many didn't believe companies would buy baled tobacco, and those like the writer quoted above just didn't think baling tobacco was the way to treat a crop that had long been handled with *pride,* a term that is frequently used to describe the proper treatment of tobacco. Marlon Waits told me that he and the men in his family with whom he raises tobacco tied their tobacco in hands until they were told that it would no longer be bought unless it was baled. Clearly some farmers did not welcome baling, and some did.

According to Extension Service professionals, this was a farmer-driven innovation, fueled by the desire to cut labor costs, but one that required the technological assistance of university research. Warehouses and tobacco companies were resistant to the move because neither was equipped to handle baled tobacco, and therefore both had to quickly adopt new equipment and procedures. I've been told that the beginning of this research was not only controversial but nearly cost some of those involved their jobs, so strong was the resistance of the tobacco companies from whom the university received research funding. In a 2001 interview for the Kentucky Oral History Commission, one former University of Kentucky faculty member referred to this period as the "baling wars" and referenced faculty getting into "hot water" for conducting research on baling. Some were so eager to package their tobacco in bales that in the late 1970s they began taking tobacco to markets in Tennessee, where bales were being accepted. One farmer told me that a few farmers used hay balers when they didn't have access to the new balers the university had designed.

The controversy was apparently not limited to adults, demonstrating just how thoroughly tobacco production permeated Kentucky community culture. Mark Roberts told me:

> Mark: That was very controversial, that was back when I was in school. And there were some farmers—their sons were in school with me and they owned warehouses and they did not like that change and, it was—they let it be known that they didn't think that I ought be doing that and ought to be part of that.
> Author: So other students let you know that?
> Mark: Oh yeah, other students. The other children.

A history published by the Burley Auction Warehouse Association (BAWA) places the resistance largely with farmers and tobacco companies rather than warehouses. BAWA concludes a brief overview of the "controversy" by stating, "Gone were the nostalgic wooden baskets filled with carefully hand-tied tobacco leaves. In their place were row upon row of hundred-pound bales. The new look seemed more efficient, and it was certainly less costly for growers and warehousemen. But an era had definitely ended."[21] This sense of nostalgia pervades many discussions of tying hands, even for those who would never go back.

Ironically, although farmers certainly benefited greatly from the move to bales, the warehouses and companies benefited even more. The amount of labor that was saved unloading seventy- to ninety-pound bales versus packing baskets with individual hands that weighed less than a pound was substantial. According to George Duncan, warehouse labor needs were cut in half. By the mid-'80s, a system for dealing with bales had become the standard that continues today, although five- to six-hundred-pound bales are now becoming the norm, as I will discuss. Six to eight bales, depending on the warehouse and the year, are unloaded onto a cardboard *slip sheet* and then bound together with metal bands (a practice that was established as the standard in the mid-1990s) to become a *pile* of tobacco, still weighing close to seven hundred pounds, that is sold as a unit just as the basket of tobacco had once been. Moving this pile onto a semitrailer with a forklift after the sale, rather than repacking hands of tobacco into a hogshead, proved tremendously laborsaving for the tobacco companies. For a number of years, the top bale was split open for inspection at the warehouse. This practice was eventually abandoned as new technologies were introduced to test moisture levels and as companies developed tracking systems for linking each bale back to a particular grower long after it left the warehouse. Today, if problems such as nesting or inferior quality are found with a farmer's bales after they reach the redrying facility, that farmer's future ability to contract with that particular company will be placed in jeopardy.[22]

The technology for baling what became the conventional seventy- to ninety-pound bale of burley tobacco has not changed since baling was introduced. A *baling box* is actually a set of connected wooden boxes (some were made of metal for a while), with a compression device such as an air cylinder or a car jack that is moved between the boxes to compress the bales. The number of boxes varies by the number of grades being stripped, which means that three is fairly standard, although many are a single box. Many farmers built their baling boxes based on plans made available by the University of Kentucky, while others like Martin bought their baling boxes from the local high school's chapter of the Future Farmers of America, which made them as fundraising projects, or from local entrepreneurs. One such entrepreneur, Maurice Corn of Anderson County, estimates that he sold between eleven and twelve thousand baling boxes over the years, including many that he made completely by hand and, later, others that he finished from pieces manufactured elsewhere.

The process of stripping tobacco changed little with the introduction of baling. However, as Roger Quarles pointed out to me, baling introduced much greater diversity in stripping-room practices than had existed with hand tying: "Then when we went to the bale package, you didn't have to—well it was just up to you really how you got it into the bale package, whether you left it in your hand or whether you just, shucked it off and put it in sort of an organized pile and then put your arms around a larger bunch and took it over to the, to the bale box, which was again compressed, to make a sturdy package." This is just one example among many of a greater diversity in on-farm practices. Of course, farmers have always had their own ways of doing things, but at one time their practices were much more alike than they are today.

In the stripping rooms I worked in or visited, tobacco is still stripped in the manner described above and then neatly laid in a pile on the stripping bench rather than held in a hand, stems pointing toward the person stripping. The piles are then placed, by grade, either into an intermediary box or directly into the baling box that holds the appropriate grade. The intermediary boxes are placed a couple of feet from the bench so that the person stripping the leaves can pick up a pile as it grows large and turn around and place it in the box without walking across the room. In other stripping rooms, the piles of stripped leaves are walked directly from the bench to the baling box. In either case, the piles are alternated as they are placed in the twelve-by-thirty-six-inch baling boxes, so that the stems are butted up against the sides of the box and line both ends of the finished bale. The box of tobacco is periodically compressed, and when the baling box is full to the appropriate point of twenty-four to twenty-six inches high (some have a "full" line drawn with a marker), it is compressed a final time, and the front of the baling box is opened. The bale must be tied with cotton twine, and it is then removed from the box and stacked in the barn. In large, nonconventional stripping rooms (discussed below), there may be one or more persons whose entire job is bringing tobacco in, packing the baling boxes with piles of graded tobacco, and removing the stalks and bales to the barn.

Finished bales are then stacked in the barn until they are brought to be sold. I will return to issues of pride and nostalgia around the marketing of tobacco later, but here it is important to point out that the aesthetic standards that were attached to tying hands moved to baling tobacco and still exist, although many express the opinion that the meeting of these standards varies widely in practice. Farmers continue to have strong opinions about what a good bale of tobacco should look like—the stems should be evenly lined up on the ends as though they'd been shaved off with a chainsaw, the color should be consistent throughout the bale, there should not be loose leaves sticking out—even if many can no longer attain this aesthetic goal because they do not strip and bale all or perhaps any of the tobacco themselves. Many farmers no longer strip any tobacco, while farmers like Martin still strip and bale part of their tobacco crop between other farm tasks, and others continue to do it all themselves.

The effect of appearance on the price that tobacco ultimately brings, currently or in the past, is a complicated topic. Some of the farmers I have spoken with swear that presentation was once a determining factor in price. For instance, Donald Morse, a former farmer and warehouseman, told me, "Dad was a pretty neat farmer, and back at that time neatness went a long way. Quality of course went further, but neatness was a big factor in selling your tobacco." Others have told me that the aesthetic that farmers were reaching for was not required by the tobacco companies:

> Author: I've heard that a lot of people took real pride, in their, their hands of tobacco?
>
> Maurice Corn: Oh, you better believe it. They, some people could make the messiest—you ever saw. But a lot of people really, I have seen crops where every hand was the same and, prettiest you ever saw. But they don't—the buyers wouldn't give you a bit more than that than they will the other.
>
> Author: Oh really?
>
> Maurice Corn: No.
>
> Author: So it didn't help your price?
>
> Maurice Corn: Didn't help your price but it, but a lot of people just, you know, didn't like sloppy tobacco.

Similarly, Noel Wise talked about appearance, and he used bales rather than hands as the example. I asked, "A pretty hand didn't bring a better price?" and he responded, "I never did see it, no. Tobacco companies didn't fall for that. It's hard to fool that tobacco company." He explained further: "And I have seen on a tobacco market floor, when we had, the auction system. They first started that baling and here was a person that every stem was even on the end of his bale. Boy it was a pretty thing to look at, you know he'd put a lot of time into it [*laugh*]. And I have seen these tobacco buyers come by there and they'd knock that over there, they wanna see what's under [it. They'd] say, 'That's on top now what's underneath?'"

According to Corn and Wise, neatness did not determine the price (and Wise suggests that being *too* neat raised suspicions), but farmers all thought it did or at least acted as though it did. In other words, there was a pretense that they were performing for the tobacco companies when in fact they were performing for one another. Charles Long compared this performance with his experience working in a factory for thirty-one years: "We made silicone caulk, and our plant manager said . . . 'You know all these years . . . we've been mak[ing] this guy a Cadillac and really all he wanted was a Ford.' So, I mean. Those hands didn't have to be tied just perfect.

It didn't have to lay on that basket just perfect." I asked, "So that wasn't for the tobacco companies, that you were doing that?" and he responded, "No." Instead, it was so that "they could say, 'That's my tobacco. That's mine,'" and they could say, "'See how many pounds I got per acre. See what my price was.'" Charlene Long, Charles's wife, summed the matter up: "It was pride in themselves."

Some farmers still emphasize the effect that aesthetic qualities have on the price that their tobacco brings even in the very different current climate of tobacco marketing, while others insist that it does not matter. The consistency of the color of the leaves throughout the bale is one visual aspect that certainly makes a difference because the color determines the grade put on it by the grader, and if a bale is a solid color that means it is all the same grade, and it will therefore bring a higher price. Because grading is still done visually—a grader assigning a grade to a pile of tobacco based on what he can see as it moves past him on a conveyor belt—it is hard to believe that neatness does not have some impact, if only at the extremes.

In some cases, the conventional stripping room is undergoing a range of changes, another example of the increased lack of typicality among tobacco producers. While some large farmers have multiple conventional stripping rooms going at one time, others have moved out of the stripping room altogether—stripping tobacco on the back of a wagon out in the open in the early fall or in the central corridor of a barn in order to accommodate a large crew of fifteen to twenty people (the most I've seen working at one time in a conventional stripping room is seven), moving the operation from place to place in order to be near the tobacco rather than bringing the tobacco to a central location. The greatest changes to stripping rooms, however, involve the movement from the small bales I have just described to large bales weighing five to six hundred pounds. These large bales require a new baler, as well as a much larger stripping space and a forklift or other similar equipment to move the bales around.

The push for a move to large bales is coming not from farmers but from tobacco companies. Philip Morris, in particular, is pushing for large bales because of the labor-saving benefits it will provide to them, and at one point the company offered some farmers cost-sharing opportunities to encourage them to make the switch. Many farmers, however, see large bales as a cost that will bring them little return. Although there is often the possibility of some degree of saved labor, it varies greatly by farm, and both farmers and extension agents agree that in many cases the costs outweigh the benefits. The labor savings come from less frequent compression and removal and from the loss of the need to orient the stems because tobacco leaves go into the big baler with no particular orientation, or as *tangled leaf*. The costs are many: the baling boxes themselves are expensive, with one to three needed at several thousand dollars apiece; if only one is purchased in order to save upfront costs, labor savings may be lost since two grades will have to be boxed until time to bale them, which doubles a portion of the labor since that tobacco will be walked from the stripping bench to a holding area and then from box to baler. In addition, it is necessary to have equipment with which to move the bales around and a large stripping space with a flat floor. Not only are conventional stripping rooms too small, but most, like the barns they are attached to, have earthen floors. And yet many growers are finding ways to adapt as they are gradually being forced to shift to large bales. Marlon Waits, for instance, told me in 2012 that he and his brother and cousin had devised a system for packaging in big bales without creating a new area for stripping their tobacco. They continue to strip it in their conventional stripping room and then bale it, loosely, in small bales so that they can easily move and store it. Prior to taking it to the receiving station, they break open the bales and rebale it in a large baler they have purchased and made room for in the barn. They move the large bales around with a duckbill, a type of handcart once used to move piles of tobacco around when it was tied in hands and layered onto baskets at the warehouse.

There is no information available about the percentage of farmers who have made this switch, but it appears that the move to big bales is inevitable. Many of the arguments being made for and against large bales are identical to those made about small bales in the early 1980s—such as concerns about "nesting" and moisture content—but so far the move to large bales has been much slower because for the majority of farmers the benefits are just not there. Although many farmers do not believe that they will benefit economically, they told me that eventually the manufacturers will require them to make the switch. That eventuality appears to be close: as of the 2011 crop, companies were beginning to penalize farmers at a rate of five to ten cents a pound if they did not bale their tobacco in large bales. This is widely viewed as another possible point at which the small grower will be squeezed out.

Chapter 3

Taking Tobacco to Market

The auction system developed and evolved over the centuries but was basically in place by the early twentieth century. Throughout the century, the opening of the market season was an important event marked by local festivals and widely covered by the media. For instance, in 1939 the opening of the markets in Lexington, for many years the largest burley marketing center, included "a parade, the crowning of a movie star as queen, a carnival, and French follies."[1] Year after year, a similar photograph was taken and published to commemorate the opening of sales: men in trench coats and fedora hats lined up on both sides of a row of piles of tobacco, in the midst of the first day of sales. The photo usually included the warehouseman, the auctioneer, the ticket marker, eight to ten buyers, and often a political figure such as the commissioner of agriculture or the governor, showing his support of Kentucky's largest cash crop through his presence on opening day. The sound of the chant of the tobacco auctioneer has become an iconic American sound—in part through radio advertisements such as the American Tobacco Company's, which featured the auction chant ending with "Sold American!"—and it became a focal point of nostalgia when tobacco warehouses closed their doors. R. J. Reynolds began to sponsor tobacco auctioneer contests in the 1980s.

The marketing of tobacco has of course changed significantly since the 2004 buyout, which brought with it the end of the tobacco auction. Beginning with Philip Morris's "Partnering" program in 2000, many farmers had in fact begun to contract directly with companies before the buyout. The advantage of direct contracting for farmers was that they saved the fees that were paid to the warehouse, netting them several cents a pound—a particularly attractive gain as they were coping with the loss of quota. This turn of events was controversial, however, as many growers and others on the production side feared the threat that it posed to the future of the tobacco program and to tobacco warehouses and those employed by them. Others accepted it as the next phase of tobacco production and gladly sold their tobacco without paying warehouse fees. Interviews conducted with growers and agricultural professionals as late as 2002 for the Kentucky Oral History Commission include strong statements of support for the continuation of the program. Rod Keugel, a tobacco grower and a former president and longtime board member of the Burley Co-op, said in an interview in 2001:

> The effort to destroy the tobacco program, is just unbelievable to me, I cannot believe that farmers, are willing for ten cents a pound, to contract *knowing* surely they know, there's an ulterior motive to what the companies are doing and, some of the younger ones out here, I guess I would include myself in the younger generation of farmers since our average age is about sixty, think that "because I raise good tobacco that the company'll be good to me and take care of me." And maybe—I hope they're right, but I don't think they are. It's never been true before.

Despite the uncertainty, by the time of the buyout tobacco growers generally welcomed it for a number of reasons. However, it has become increasingly clear to me that everyone involved on the production side (meaning everyone but the companies and perhaps some in the political arena) "welcomed" it primarily because, by 2004, a buyout seemed to be the only way to continue to raise tobacco with any profit at all. The quota reductions in the years leading up to the buyout were so severe that one farmer described his experience this way: "They knew what they were doing, they cut us, they kept cutting us for three years, down—I think from fifty thousand [pounds] they cut me to ten thousand. Then they said 'Well now this year we're gonna buy out.' So I got sixty or seventy thousand [dollars], and if they had had it when they should have—started talking about it, I would have had four hundred thousand [dollars]. Now that's a big—that's a big big difference."

Many growers, like this one, feel that the timing of the massive quota cuts in the years just prior to the buyout was intentionally planned to benefit the companies—particularly since a buyout was discussed throughout the period during which quota was being cut at record levels. Despite governmental oversight, the companies determined the demand and therefore had the ability not only to keep farmers guessing in terms of what would happen with quotas from year to year but also to manipulate the quotas just as this farmer described. I would argue that farmers generally welcomed the buyout because of fears induced by the roller-coaster ride they experienced in the 1990s, ending with drastically reduced quotas. The alternative seemed to be a program that would continue to shrink each year until marketing quotas were gone altogether.

I began this project in 2005, the first crop year following the buyout. At that point, those with whom I spoke were assuming a "wait and see" attitude; I was told repeatedly that "we'll see what happens." On the surface, farmers and others who told me this meant that no one knew for sure what price tobacco would bring—in itself a major change, since for over sixty years prior (the lifetimes of the majority of growers) growers could expect support prices to be announced before the opening of the market. While there was some uncertainty about whether the price would climb above the one they'd gotten the year before, and how far, they could have confidence that it would not drop substantially. What they also meant by "we'll see what happens," however, is that now they had to face the possibility that the price might drop considerably and, even more frightening, the possibility that, without the pool as their safety net, they might not be able to sell their crop at all. The price schedules that growers agree to when they sign a contract are fully dependent on company pronouncements about the grade and quality of the tobacco presented to them at the time of sale, as judged by company representatives. This leaves plenty of room for the contracting company to name the price and even refuse to buy all or part of a grower's crop of tobacco by simply grading it outside of the grades included in the contract. The contracts in fact bind neither party to buy or sell. For the first time in three generations, tobacco farmers had no guaranteed market. Many quit that first year, while others raised more tobacco, and still others raised less. Some quit and have since come back, and some have since quit.

Although there are no statistics available regarding the post-buyout population of tobacco producers, based on my fieldwork, those who "got out" of tobacco can be categorized by their circumstances. The first were farmers who were ready to retire, many of whom held on for several years only because of belief in a coming buyout that they would qualify for only if their base remained active. Roger Quarles told me that a buyout had been anticipated for many years and that many growers were "still growing tobacco because it was sorta known that, you had to be involved in the system in order to qualify for a payment if it were to occur, and so we had a good deal of people that would normally have retired, that were still raising a crop of tobacco. And so once the buyout occurred all those folks immediately stopped."

A second group were owners of quota who were leasing their tobacco to growers and were therefore only connected to the crop on paper. In addition, many "part-time farmers"—as full-time tobacco farmers call those who work off-farm jobs and grow a crop of tobacco—got out. Some farmers ceased tobacco production and increased other farm activities they were already involved in, particularly beef cattle. Some farmers have also successfully moved to niche crops and agribusiness opportunities. However, they are in the minority, despite public discourses that seem to suggest that they are the norm. Those who continue to raise tobacco are largely "tobacco men," as well as a few women, with a particular relationship to the crop and heavy financial investments in its production. According to Roger Quarles: "You know, in my situation, you know, you sit back and look 'Well what am I gonna do with these acres? If I don't grow tobacco what can I do with it?' And I've also got these barns that I've, still paying mortgages on, that still have to be maintained, still have to carry insurance, still have to paint them and all that. What am I gonna do with them? Do I just let them sit idle?"

Prior to the opening of the market for the 2007 crop, Marlon Waits explained the post-buyout era this way:

> Marlon: As long as it was pretty decent-quality crop you had a support price. If the companies didn't want it you still, you were guaranteed so much. Now you have no guarantee. Either they want it or they don't want it.
>
> Author: So does it feel . . . ?
>
> Marlon: Yeah it's a little nervy taking it in there and you don't know whether they're gonna buy it or not. So far they've been pretty good about it—of course, we haven't had what you call a "bad crop" yet.

His statement about not yet having a "bad crop" was a direct reference to the upcoming marketing season, during which farmers feared they would be delivering a "bad crop" for the first time since the buyout due to the drought conditions, as well as a late freeze in the spring of 2007.

Although many growers continued to express uncertainty about the future, most told me that things would be much worse if the buyout had not happened; some told me that they'd prefer to have the program back in place. It is clear that the new environment for American tobacco production has not shaken out yet—this is an ongoing transition. Each crop year since 2007, more growers have in fact seen their tobacco refused by the companies at sale time, fewer farmers have been offered contracts, and more farmers have quit raising tobacco altogether.

Although a few burley warehouses remain in operation in Kentucky (there were three in the winter of 2007–2008), still conducting small-scale auctions, almost all burley producers now sign a contract with a tobacco company—many larger growers sign contracts with multiple companies—for a specified number of pounds in particular grades at a price schedule determined by the tobacco company prior to the beginning of the growing season. Until 2009, contracts included various incentive programs, such as eight to ten cents more per pound if a grower produced the poundage he or she contracted to raise. Philip Morris now controls an estimated 70 percent or more of burley tobacco contracts.

The tobacco check, for generations, was an important source of cash that came at the end of the tobacco season as compensation for the year's tobacco work, and its importance is reflected in stories about bad years. John van Willigen and Susan C. Eastwood note that farmers "tell stories from their fathers and grandfathers about selling tobacco for less than the cost of taking the product to market."[2] W. F. Axton, too, relates such stories, including one about "one small planter, who had netted $325.00 for his 1919 crop, [and] came away in 1921 with just $2.75."[3] I was told a number of variants of such stories during my fieldwork, some of which were about particular people (usually family members), while others concerned unnamed farmers. One such story appears to have long been in wide circulation, as I have encountered it in both oral and print sources.[4] In this story, a farmer brought his tobacco to the warehouse only to have it bring less than the fees he had to pay to sell it. The warehouseman told him to bring him a chicken to cover what he owed. So the next day, the farmer came back to the warehouse with two chickens, and when the warehouseman asked him why he had brought two, he replied that he had another load of tobacco to sell. All of the versions of the chicken story I heard or read have a central element of the farmer as the buffoon (although an alternative reading might also see a critique of the warehouseman rather than the farmer). This story is sometimes told as a legend, with the teller explicitly saying that he wasn't sure whether it was true or not. At other times it included a clear punch line and was followed by laughter that indicated disbelief, suggesting the teller was presenting it as a joke. I was told other jokes with similar messages, including one in which a watermelon farmer deals with the low prices he's getting for his watermelons by growing more watermelon (suggesting what tobacco farmers have done and are doing today) and one in which a farmer wins the lottery and when asked what he's going to do with the winnings replies, "I reckon I'll just go home and farm until I run out of money."

While some of these examples are told as jokes, similar stories are told in a more serious tone. For instance, Wilbert Perkins told me a story about his father fitting his entire crop in the backseat of the car one year and not being able to pay off a loan of one hundred dollars that he'd taken from his uncle because he only got fifty dollars for his crop.[5] A second story was told to me on two occasions at the 2007 Kentucky Folklife Festival by Charles Long (who also told me the chicken story on another occasion), once informally and later the same day during an on-stage interview about tobacco farming. Sometime in the 1930s or '40s—the implication being preprogram—a farmer went to sell his tobacco.[6] He didn't make any money once he'd paid his warehouse fees, so he couldn't pay the grocer what he owed from the previous year. So he sold a cow, but he only made a nickel on it. He was given a check for five cents, and he brought it to the grocer to show him that he couldn't pay him, and the grocer was so taken aback by a check for five cents that he bought the check from him. In the version Long told me on stage, the grocer gave him the nickel for the check and put the check on display under the glass top of his check-out counter. This story further develops the narrative of low tobacco prices by showing that even the diversified farmer couldn't make it because the bottom had also fallen out of the cattle market. The second version, in which the grocer puts the check on display, represents what may have been a common practice related more directly to tobacco: the preservation and display of tobacco checks and receipts that reflect dismal prices. Van Willigen and Eastwood include an image in their book *Tobacco Culture* of a check written to a tobacco farmer for ten cents in 1921, and warehouseman Jerry Rankin showed me a framed 1938 floor sheet (a receipt for auctioned tobacco) hanging in his tobacco warehouse in Danville. Stories such as these argue that there was a drastic need for a solution to the problems of overproduction and the control of the industry by a handful of large corporations, and such stories may serve to justify the renewal of government monitoring through the creation of the tobacco program by arguing that farmers could no longer be expected to fight the giant companies on their own.[7] Told in the years immediately following the end of the tobacco program, these stories may express fears that the return of the free market will mean a return to such marketing circumstances.

The burley tobacco sales season opens in November and remains open until farmers are finished selling tobacco. While this end date depends on the weather and when farmers are able to finish stripping, it generally comes sometime in February, with a break at Christmas. The marketing season once opened up additional employment possibilities for farmers who were able to finish stripping their tobacco early. At one time, of course, having a truck and trailers with which to haul tobacco to market was rare, and many farmers made additional money trucking tobacco to market. Other farmers worked at one of their local warehouses during the season, as warehouses depended on seasonal help. Several farmers told me that their first job off the farm, as young men, was packing baskets at a tobacco warehouse.

Delivering tobacco to the warehouse often meant sitting in a line of trucks that stretched down the street for hours; stories are told of being in line all day or even for multiple days. Trucks loaded with tobacco would sometimes be left parked in a line overnight, security guards hired to watch them. When a farmer's turn came, his or her tobacco would be packed onto baskets, weighed, and ticketed, and a date would be set for it to be auctioned. Once the tobacco was at the warehouse, the warehouse assumed responsibility for it. It would later be graded by a government grader. The sale date was important information because farmers never missed the sale of their tobacco. Sometimes the whole family would come back for the sale, often standing by the tobacco to remind the buyers that the family depended on their tobacco income, in hopes of bringing a higher price. I heard many times that fifths of whiskey, country hams, and young pretty daughters were placed on top of many a pile of tobacco as encouragement for a good price.

Families didn't always come to the auction, however, and warehouses provided some tobacco men with time away from farm and family—time in which to drink and play cards and in some cases to lose their hard-earned tobacco money before the piles had even been sold. The warehouse business was highly competitive, and warehousemen used a host of solicitation techniques to bring in farmers. Handing out pints of liquor to every farmer who wanted one was standard at one time, and warehouses often gave away meal tickets for use in local restaurants (and I've been told that the owners of the warehouses were sometimes also the owners of the restaurants). By the end of the warehouse system, warehouses had begun to give out ball caps with their logos printed on them instead of alcohol. Warehouses hired solicitors who would make farm visits throughout the year, drumming up business for the coming season.

The relationship between farmers and warehousemen was a complex one, but generally farmers saw warehousemen as filling a role that was predominantly helpful to them. It was in the interest of the warehouseman to keep a farmer happy since he could decide to take his tobacco elsewhere in the days when there were multiple warehouses in most tobacco marketing centers; for instance, in 1968 there were 245 burley tobacco warehouses in Kentucky alone.[8] A warehouseman might provide encouragement for buyers to pay a good price for a particular farmer's tobacco; he might repack tobacco so that it looked its best; or he might even buy it himself to resell later, sometimes at a loss. According to former warehouseman Ben Crain, describing what happened when the house bought some tobacco, "On occasion you will make some money on a pile of tobacco. Or, what you really hope to do is to break even. And satisfy that customer." There were advantages to moving to a new warehouse every now and then. According to Roger Perkins, "You'd get some good deals the first year or two" because "you know they'd take care of you, I mean they'd make sure you got a good sale. I mean just like anything else, that warehouseman could make a lot of difference back in those days." He went on to explain that "they'd buy a pile or two of your tobacco for, a dollar or two you know above what the buyers were paying. After all, it's all about the bragging rights of what the average says on the ticket so I mean you know." At the same time, farmers obviously knew that warehousemen were profiting from the fees that they paid them to have their tobacco sold, so they never felt that they were on equal footing. Everyone employed in the warehouse system—operators or warehousemen, auctioneers, ticket markers, office staff, and other warehouse employees—lost their jobs as warehouses were forced to close. Towns across tobacco country continue to struggle with what do with these massive empty buildings; many can't afford to tear them down, and as they age, safety concerns grow.

I had been told many times that a tobacco warehouse is "the coldest place on earth," and I experienced that cold firsthand when I attended a sale day at one of three Kentucky burley tobacco warehouses struggling to remain in operation during the 2007–2008 marketing season—on a very small scale: Farmers Tobacco Warehouse No. 1 in Danville.[9] Ben Crain, former R. J. Reynolds buyer and Lexington warehouseman, offered to take me to the sale, which he participated in as a buyer. He told me that this was the closest thing to a "good old-fashioned auction" that I'd be able to see. Indeed, the major differences between this auction and the many descriptions that I have heard were primarily in the number of participants, the lack of a government grader, and the speed at which the sales were made. Of course, all of this made for a very different atmosphere than that of past auctions as they were described to me.

The warehouse was full of rows of piles of tobacco that had been delivered over the preceding days, each with a list of the farmers whose tobacco was in the row. Each pile had a ticket on it, and the picking up of this ticket signaled the start of the sale of that pile. The participants on this day included the warehouseman, the auctioneer, two other warehouse employees on one side of the row of piles, and four buyers across from them. There were a number of growers there—presumably everyone who had tobacco being auctioned—following along behind the warehouseman and the auctioneer. As the group approached each pile of tobacco, its ticket was picked up by a warehouse employee and handed to the warehouseman, who started the bidding process. The auctioneer took up the opening price with his chant, and then the buyers placed their bids in a mixture of gestures and words that were mostly incomprehensible to me. The auctioneer announced the final price in abbreviated form (he'd say, for instance, "fifty-eight" instead of a "dollar fifty-eight"); wrote the grade, the price, and the symbol of the buyer on the ticket; and dropped it back on the pile. This all happened very quickly, although not nearly as quickly as at the auctions in the days before the buyout, when five or six hundred piles of tobacco would be sold per hour by one auctioneer. According to former warehouseman Ben Crain, in those days a sale would occasionally be stopped so that the group could return to look again at a pile that had been sold because a buyer might say, "That pile that I bought back there is following me." But stopping the sale was avoided whenever possible so that the rhythm would not be interrupted.

During this sale, I followed along with the small group of growers. We trailed behind, looking at the marked tickets once the corresponding pile had been sold—sometimes because the tobacco was theirs but mostly just to see how it was selling—and we compared the tickets and discussed the fairness of the prices that the tobacco was bringing. One grower I spoke with represents one type of circumstance that brings growers to sell their tobacco at auction now. He had raised about ten thousand pounds (about four or five acres), which was about three thousand pounds more than he had contracted to sell to a tobacco company. Because the companies were not buying much over contract this year, he needed to sell his three thousand pounds of *carryover* somehow, and the auction was one of the options that he had to choose from. *Carryover* is a term left over from the days of the quota system; under the tobacco program, if you ended up growing more pounds than your quota permitted, you could carry a small amount over to the next year. This grower was considering selling all of his tobacco at the warehouse the following year rather than contracting with a manufacturer if he did well at this sale.

The buyers at this sale reflected another great difference from the classic tobacco auctions in that they were neither buyers for large tobacco companies nor leaf buyers for multiple companies worldwide. Instead, they were primarily buying for large growers who were otherwise not going to meet their contract poundage and would therefore lose incentives, for the Burley Co-op (which bought about two million pounds of the 2007 crop, with plans to establish marketing relationships in China), and for the *house* (meaning Farmers Tobacco Warehouse No. 1). The prices ranged from $1.25 to $1.62 a pound on this day, with most of the tobacco in the high $1.50 range. This was consistent with the prices that growers were being paid on contract. However, those selling their tobacco at auction would not receive incentives of up to ten cents a pound for fulfilling their contracts, and they had to pay a warehouse fee. The warehouse-man Jerry Rankin later told me that a number of the farmers who were selling their carryover were selling their least desirable tobacco through the warehouse, or they were selling tobacco that was not graded and baled in a manner that *showed* well. Jerry buys much of the tobacco that sells through his warehouse himself, and then he repacks it so that it shows well and meets the needs of the companies to which he hopes to sell it. He expected to see about three million pounds of burley go through his warehouse that season, less than half of the approximately eight million pounds that would have gone through his warehouse in a good year in earlier times. He was taking a considerable risk since he had no guaranteed buyers. Jerry told me in an interview, "What I'm doing at the warehouse right now is pretty much spinning my wheels. I'm offering a service I guess." At the time of our interview, the warehouse had only made a profit in two of the four years since contracting began, and it had either broken even or operated in the red in the other years.

Of course, very little tobacco is sold at auction today. Most is brought to receiving stations set up by individual companies by growers who have signed contracts, and these are very different places from warehouses. Martin and I had planned that I would go with him to sell his tobacco, finishing off the year, but in the end the timing did not work out, and I was unable to accompany him on any of his three trips. I did go along with the Waits family (brothers Marlon and Robert and their cousin Carl, who have taken over the family farm operation from their fathers, Ray and Harold) to the Philip Morris receiving station in Frankfort in mid-December, on one of several deliveries that they made.[10] Under the contract system, farmers are given precise appointment times for the delivery of their tobacco, and they can deliver only a specified amount each time. Many growers appreciate that they no longer have to wait in long lines to deliver their tobacco and that they are handed a check on the same day, but they don't like the loss of their ability to choose when to deliver their tobacco—especially if circumstances such as rain arise. Some also express regret at the loss of personalized connections to the sale of their tobacco. Rather than a warehouse owned and staffed by local people, Philip Morris receiving stations employ a combination of local workers and company representatives from North Carolina and/or Virginia.[11]

Although I've been told that generally there is little or no wait time at the receiving stations, we did wait a couple of hours on this particular December day. The Waits family had three pickup truckloads and one very large trailer. Trucks drive directly into the building through four large overhead doors that lead to four separate unloading areas. The bales of tobacco were unloaded onto large cardboard slip sheets in stacks of six bales of the same grade of tobacco, which were then banded together with three metal bands about one inch wide, making a pile or *basket* of tobacco that would move on to the graders. Marlon had told me numerous times that his family likes to unload their tobacco themselves because the Philip Morris employees who unload it handle it too roughly, tossing it off the truck in a manner that could hardly be more different from the days of carefully crafted baskets of tobacco tied in hands. I watched that day as the Waits family managed the unloading so that their tobacco was treated with care: Carl climbed up onto the truck and handed each bale down to Robert, Marlon, and the Philip Morris workers; Robert and Marlon arranged the bales to ensure that their high quality showed.

Forklifts whizzed back and forth, moving the baskets first to a holding area until it was all unloaded and ready to be sold, and when their turn came their tobacco was moved onto a conveyor belt where the moisture level of each pile was tested, the tobacco was weighed, and it was graded by a Philip Morris grader. The grader who handled the Waitses' tobacco, whom I had met while we waited, invited me to stand with him on his elevated platform above the conveyor as he graded their tobacco. I was told that I could not take photographs during this process. The grading system used by the tobacco companies is much more complex than the grades into which tobacco is stripped, resulting in three types of grades applied to each pile. They start with the four grades that farmers use (trash, leaf, lugs or cutters, tips), and then add a numerical grade for quality, followed by an "internal grade" that is based on color and determines the use to which particular tobacco will be put.

Once tobacco is sold, it goes to a *redryer* to be processed, and then it is allowed to age for three years until it goes to its final destination, a manufacturing plant. With the sale of their tobacco, farmers come to the end of the crop year. For most farmers, this end didn't come until January or even February 2008. Although I wasn't able to make the trip with Martin to sell his tobacco, my last day of fieldwork pertaining to the 2007 crop year, in February 2008, was also his last day stripping tobacco. Somehow this seemed appropriate, as did the fact that, as I interviewed him once again that day, winter calves were being born, and as I drove off he was bottle-feeding a newborn calf.

I later learned that he received good average prices on his 2007 crop and that he not only reached his contract weight of thirty thousand pounds, therefore receiving an additional nine cents a pound, but also had about twelve hundred pounds of carryover that he was able to sell to another farmer who needed it in order to meet the pounds he had contracted for. What had looked like a disastrous crop back in the summer turned out, for Martin, to be decent. Martin and I discussed the 2007 crop in an interview just before I left in February, and he said, in part:

Martin: I was hoping that I would at least get my money back, and I'm gonna do that without any trouble. . . . That way it makes you feel good enough to start another crop [*both laugh*].

Author: As long as you get your money back you can start another crop.

Martin: Yeah you gotta get your money back [*both still laughing*]. My daddy always said, "You put your money in your crop all year, and then you get it back in one lump sum, and then you can do something with it." You can do a whole lot more with several thousand dollars than you can, a hundred dollars here and a hundred dollars there.

Author: Like start another crop huh?

Martin: Yeah, like start another crop. But I reckon, I reckon [this year] I'll raise about what I tried to raise last year. Of course, no two crops are the same. I've been raising tobacco since I was nine years old and, and never had two crops the same. It's always different.

Tobacco growers must be able to adapt their knowledge and skills to changing circumstances. Despite commonly held stereotypes of farmers as stubborn and old-fashioned, tobacco farmers have in fact proven to be quite willing to adopt new innovations. Farmers adopted MH-30 ("sucker dope") in the 1960s, baling in the early 1980s, and float beds in the 1990s. Today those who continue to raise tobacco are adapting to a free-market environment. All of these changes, and others, represent rapid and substantial technological and marketing changes. Tobacco farmers often told me that when a farmer sees that a new technology works well, is accessible and affordable, and will improve his tobacco income, he very likely will adopt it. As I will discuss in later chapters, many farmers express nostalgia for a time when the work was done differently—but they would not choose to return to those times. Their nostalgia is for the changed meanings of the work, rather than for the work itself. For at the same time that the changes in technology and marketing practices that I have described were happening, the political and social meanings of tobacco were also changing dramatically.

Part 2

The Shifting Meanings of Tobacco

Introduction to Part 2

In my early fieldwork with tobacco producers I learned that tobacco is most often talked about in terms of change. Through my ethnographic research I learned about a range of changes that have taken place on tobacco farms—from technological innovations to new marketing practices. But other kinds of changes pervaded my fieldwork: the changed political and social meanings of tobacco, the effects of which were raised both directly and indirectly by farmers, extension agents, and others. My recognition on that day in the summer of 2007 that tobacco was missing from the walls of the Kentucky Department of Agriculture led me on a search for when and how the crop that had until very recently been the state's largest cash crop had disappeared. I now turn to the shifting political landscape of tobacco production through the analysis of a particular rhetorical site, the newsletters of the Kentucky Department of Agriculture. Kenneth Burke has defined rhetoric as "the use of words by human agents to form attitudes or to induce actions in other human agents."[1] In other words, rhetoric is persuasive communication, and rhetorical analysis is the process of examining a particular act of communication in order to understand what attitudes and actions an audience is being persuaded toward. When viewed as a location from which the KDA purposefully communicates to the public, the newsletters become a rhetorical site that can be examined in order to understand the attitudes the KDA has hoped to form about Kentucky agriculture more generally and tobacco specifically and the actions it has attempted to induce in farmers and others. The KDA's rhetoric is of course not static; rather, it has shifted over time along with the political meanings of tobacco. Such shifts are the primary focus of the following two chapters.

In these chapters, I examine the newsletter of the Kentucky Department of Agriculture from when it began in the 1940s through when it became an online publication in 2008, a period of dramatic changes in tobacco. The mid-twentieth century saw the rise of public awareness of the health effects of tobacco use, the increased importation of tobacco, and the beginning of the decline of American tobacco production. Later in the twentieth century and into the twenty-first, the evidence of the dangers of tobacco use was increasingly accepted by the public, taxes on tobacco products increased, lawsuits against tobacco companies were finally successful, and the federal tobacco program ended. My rhetorical analysis of the newsletter as it reported and commented on these and other events and issues demonstrates the shifting meanings of tobacco as interpreted by the state. When the newsletter began in the 1940s, tobacco was understood as an important commodity; rising smoking rates following the First World War were good for the Kentucky economy. With mounting attention to the health effects of smoking from the 1950s onward, the KDA sought ways to defend this important cash crop. "King Tobacco" was celebrated as central to the Kentucky economy and later, as the threats became increasingly serious, as the centerpiece of Kentucky's economic heritage. By the 1970s the rhetoric had shifted from celebration to more direct responses to the "attacks" on tobacco. "Heritage" came into use in reference to tobacco's symbolic rather than economic importance, and by the late 1990s tobacco was presented as a way of life that was in the past. By the time the federal tobacco program ended in 2004, tobacco had disappeared almost entirely from the pages of the KDA newsletter, replaced by an agricultural economy based on healthy crops.

There are a number of volumes that probe the history of the tobacco industry, including important recent studies by Peter Benson and Allan M. Brandt.[2] Such studies detail how tobacco companies "created a set of powerful rationalizations for denying the harms produced by tobacco and sustaining the financial success of the industry" over the course of the twentieth century.[3] The scrutiny of industry practices—from hiding knowledge of tobacco addiction and disease for decades, to unabashedly marketing to particularly vulnerable populations—has proven Big Tobacco to be perhaps the most nefarious of all industries.

Benson reminds us that the social meanings of tobacco "did not change with one fell swoop."[4] "Tobacco companies," Benson continues, "never simply responded to a problem that existed apart from their involvement in shaping what the problem looked like exactly. The tobacco problem was constructed out of the dialectical relationship between the intensified criticism of the industry in the last half century and the responses and justifications provided by the industry."[5] As my close reading of the newsletter makes clear, the state rhetorically aligned the economic health of the state with that of the tobacco industry. As events and circumstances changed for (and were changed by) the industry, the rhetoric of the KDA shifted as well, often mirroring the public rhetoric of the industry. It is hardly surprising that the second-largest tobacco-producing state aligned itself with the tobacco industry at one time. For instance, it is well known that congressmen from tobacco states were among the strongest advocates for the industry. However, just how the state publicly aligned itself with the industry has not been examined, nor have the attempts to rhetorically disentangle the state once the tides had finally turned against the industry and the state had begun to realign itself with a diversified, *healthy* agricultural economy—one that it is working to create even today.

The Kentucky Department of Agriculture Newsletters

According to Commissioner of Agriculture Ben S. Adams in April 1955, "No other enterprise in our State has as great an influence on the entire economy of Kentucky as the tobacco industry. Naturally, whenever the well-being of the tobacco industry is in jeopardy it is a matter of concern to everyone."[6] In January 2004, Commissioner Richie Farmer wrote, "There can be no doubt that Kentucky's agriculture community faces many challenges. Tobacco farmers are moving to diversify their farm operations."[7] While the contrast between these two statements most obviously reflects the decline in tobacco production over the decades, it also reflects a markedly changed rhetoric.

The KDA newsletter is the focus of these chapters for a number of reasons. Foremost, while multiple state agencies within the executive branch of state government, along with the legislative branch, are involved in the creation and implementation of agriculture policy, the Kentucky Department of Agriculture is the central state government agency responsible for "promoting Kentucky agriculture in all its aspects."[8] The agency was first established in 1876 by the Kentucky legislature as the Bureau of Agriculture, Horticulture, and Statistics. Although the governor initially appointed the commissioner of agriculture, the Kentucky constitution of 1891—which remains in place today—established the position as an elected one.[9]

The department provides "regulatory, service, and promotional programs and services" related to Kentucky agriculture,[10] which means in part that the KDA serves as the principal political voice of Kentucky agriculture to the citizens of Kentucky and beyond. While an analysis of the KDA's internal documents would be revealing in terms of the department's actions and motivations, the newsletter reveals something different: the agency's intentional communications to the public. The newsletters are a rhetorical site through which the KDA attempts to shape attitudes and induce action. In other words, they are a site through which the state presents Kentucky agriculture not just as it *is* but as the state wants it *to be.*

The newsletter of the Kentucky Department of Agriculture[11] was originally published by the Division of Markets, an office "created by the 1940 Kentucky General Assembly as the unit of the Department of Agriculture, Labor, and Statistics responsible for promoting the efficient marketing of Kentucky's agricultural products."[12] The Division of Markets began publication of the newsletter in 1941, presumably as one means of satisfying this responsibility. The newsletter provides an account of events and issues of interest to the agency and its audiences, reporting on the goings-on at the agency and the important agricultural issues of the day. The department is also charged with a number of regulatory duties not specific to agriculture (such as testing retail fuel pumps), and therefore such issues also appear in the pages of the newsletter. However, the newsletter must be read with the purpose in mind of "promoting the efficient marketing of Kentucky's agricultural products," as a reminder that the KDA does not simply report news. Understanding the newsletter as a rhetorical site includes a consideration of what the KDA chose to report on and what it chose to leave out, language and word choices, the use and placement of images, the attribution of authorship, and how all of these things changed over time. Of course, the Kentucky Department of Agriculture cannot itself speak. The employees of this government agency speak on its behalf, but the voice of the agency is implied through the lack of attribution to any one author in the majority of the newsletter's articles. There are significant exceptions, however, such as an ongoing feature long called the "Commissioner's Corner," in which the current commissioner of agriculture provided editorials on current events; guest columns by farm organization and government leaders; and occasional articles reprinted from other publications.

The publication of the newsletter began during the era of "new agriculture" in America, during which the government was becoming irreversibly involved in agriculture, and farmers were actively encouraged to farm in new ways.[13] The newsletter itself, as well as particular events, programs, and policies reported on within its pages, must be viewed within this context of increasing efforts to replace vernacular knowledge with research-based knowledge created and disseminated by official sources.[14] Of course, attempts to professionalize agriculture were far from new. As early as the 1780s, agricultural societies had formed in the United States, "in part imitating earlier English organizations, committed to seeking out and encouraging better and more productive ways of farming."[15] In 1862, the US Department of Agriculture (USDA) was elevated to an independent governmental agency, and the land-grant college system was created through federal legislation, resulting in the establishment of land-grant colleges in every state that would, in part, provide agriculture education.[16] The first decades of the twentieth century, however, brought increased governmental intervention in agriculture, most notably through the creation of the Extension Service in 1914 by the Smith-Lever Act[17] and later through the enactment of New Deal legislation that created federal commodity programs, including the tobacco program. With the New Deal legislation, "while nobody realized it yet, the government was in the agricultural market to stay."[18]

The Second World War years have been called a "productivity revolution," with improved crop and animal varieties, mechanization, and the introduction of petroleum-based chemicals.[19] This new agriculture of the 1940s—still with us today—ensured that as federal and state governments increased efforts to assist farmers, they were also increasing their power over agricultural knowledge and practices and therefore over farmers as well. Through agricultural programs and the "productivity revolution," the state claimed official science-based agricultural knowledge as superior to the vernacular knowledge of farmers. The newsletter of the KDA, with its beginnings in the early 1940s, can be seen as one means through which the KDA attempted to claim its role as both an advocate and a source of information for farmers. As it meted out information, it was also working to "induce action" in farmers.

The KDA appeals to its farmer audience most obviously through the provision of articles about such topics as variety trials, the testing of farming practices and technologies, and advice for farm women.[20] Yet farmers generally look not to the KDA for their pedagogical needs, but to their farmer neighbors, to the Extension Service, to commercial sources such as farm-supply stores and seed and equipment companies, and increasingly to tobacco companies. Therefore the pedagogical articles in the newsletter also serve the purpose of arguing that the KDA knows farmers and understands what is important to them and that the KDA has the authority to impart this knowledge. Other articles—such as descriptions of basic farming practices—serve pedagogical purposes for a nonfarmer audience, while simultaneously seeking identification with farmers by showing them that the department knows what farmers know. According to Burke, "Here is perhaps the simplest case of persuasion. You persuade a man only insofar as you can talk his language by speech, gesture, tonality, order, image, attitude, idea, *identifying* your ways with his."[21] In other words, by mirroring farmers' knowledge back to them, the KDA is not only claiming knowledge but also attempting to demonstrate that the interests of farmers and of the agency are aligned.

Another way in which the newsletter can be seen attempting to shape attitudes and induce action is through an ongoing presentation of *who we are,* with the "we" extending from farmers to Kentucky agriculture to the state as a whole. The reader is asked to visualize him- or herself in the story being told through specific strategies such as pronoun usage (directly addressing the reader as "you," for instance), as well as through feature stories about successful farmers and celebrations of particular crops and events. Not only do such pieces celebrate *who we are,* but they also attempt to persuade the audience to action toward *who we want to become.*[22] The KDA's presentation of *who we are* changed over the years, reflecting shifting perceptions of how audience(s) should perceive Kentucky agriculture and what Kentucky agriculture should become.

Over the years, tobacco production as represented in the newsletter shifted from *self-evident* to *self-conscious,* each signifying a different attitude.[23] *Self-evident* coverage presents tobacco as a farming tradition that was accepted as "part and parcel of a way of life"—in Barbara Kirshenblatt-Gimblett's words[24]—and includes those articles that report on ongoing activities that are routine aspects of tobacco production and marketing. In contrast, *self-conscious* coverage calls attention to tobacco in terms of its symbolic importance as tradition and/or heritage, makes claims about tobacco's centrality in

Kentucky life and culture, and celebrates tobacco's role in the past and the present. The self-conscious coverage differentiates tobacco from other commodities in the pages of the newsletter. For instance, market reports were a regular feature of the KDA newsletter for decades, and tobacco was reported on as one of the many crops produced on Kentucky farms. This represents self-evident tobacco coverage because attention is not being called to the symbolic value of the crop; rather, it is included in a routine report of the markets. Another example of self-evident coverage is regular coverage of the pressing tobacco policy issues of the day. This example is particularly important because this coverage faded in the late 1990s, a time in which some of the most significant tobacco policy changes were taking place. On the other hand, a pictorial series about the steps farmers take each year in order to get tobacco—and no other commodity—from seed to market is an unmistakable expression of symbolic meaning. Such displays and celebrations represent self-consciousness about the importance of tobacco to the Kentucky economy and culture, a self-consciousness lacking from the treatment of all other farm products.

Self-evident and self-conscious coverage of tobacco long existed side by side. Tobacco was not fully "self-evident" at any point in the twentieth century, certainly not at the time the newsletter began. As evidenced by the writings of William Tatham in 1800, for instance, tobacco may indeed not ever have been fully self-evident for Euro-Americans. There was great self-consciousness about the crop in the 1940s, based in large part on the events that had taken place in the preceding decades—the Black Patch Wars; the rise, fall, and reorganization of cooperative associations; and the creation of the tobacco program. Such events made full self-evident status impossible. Yet the self-consciousness of the 1940s was about the economic value of the crop, not about the symbolic importance of tobacco to Kentucky.

The self-conscious coverage of tobacco climaxed in a rhetoric of tobacco as "heritage" that was deployed by both the industry and the KDA as a defense of continued tobacco production. According to historian and curator Sir Roy Strong, "It is in times of danger, either from without or from within, that we become deeply conscious of our heritage."[25] The labeling of aspects of culture as "heritage" most often takes place in periods of "social transformation"[26] and can be seen as "a mode of cultural production that gives the disappearing and gone a second life as an exhibit of itself."[27]

Tobacco production was perceived as threatened since the beginning of Euro-American production, as the metahistory provided in this volume demonstrates. Such threats have taken different forms in different periods—overproduction, undervaluation, fluctuating markets, company consolidation, farmer cooperatives, and finally "health scares," falling smoking rates, foreign-grown tobacco, and taxes. The KDA newsletter provides a particular perspective on the shifting threats to tobacco, including where threats have come from and, perhaps most important, the political responses to these threats—including the deployment of a rhetoric of heritage. Initially, *tobacco is our economic heritage* was used as a means of justifying continued production in the present. Gradually, however, as the political status of tobacco was reversed, heritage and economics were discursively separated until tobacco *heritage* came to refer to a way of life of the past, even as farmers continued to depend on tobacco for their livelihoods. There is a common perception in the general public within and outside of Kentucky that tobacco is gone; I argue that this is in part a result of the rhetoric of tobacco as heritage.

In the following two chapters, I describe the coverage of tobacco in the newsletter of the KDA from its beginning in 1941 through April 2008. I examine representative articles and trends in order to demonstrate the shifting political symbolism of tobacco as reflected in the pages of the newsletter. Coverage of tobacco in these newsletters is in part seasonal, as over the years there is more coverage of tobacco during the harvest and sales seasons and when there are events that take place that center on tobacco. The August through December issues of the KDA newsletter generally include more tobacco-related content than either the spring or summer issues because this is the season during which the KDA was most active in its former role in tobacco marketing, as it was responsible for such things as testing warehouse scales and producing daily market reports. Although the KDA newsletters covered issues specific to dark air- and fire-cured tobaccos (the other major types of tobacco grown in Kentucky) along with burley, there is much more attention paid to burley. Historically, dark tobaccos accounted for only about 5–7 percent of Kentucky tobacco income,[28] resulting in burley's status as a regional symbol.[29]

The following two chapters are structured around distinct eras of tobacco coverage, differentiated by the changing rhetorical strategies used by the KDA in its presentation of tobacco and related topics as the political symbolism of tobacco shifted. Chapter 4 begins in the 1940s and ends with the late 1960s, a period in which the coverage of tobacco in the newsletter moved from the treatment of tobacco as a symbolically self-evident farm commodity that was highly valued despite the tumultuous character of the industry, to one valorized and defended as essential to the Kentucky economy. Chapter 5 begins with the shift to an overt rhetoric of tobacco heritage and traces the erasure of tobacco almost entirely from the pages of the KDA newsletter even as the crop continued to contribute to the Kentucky farm economy and to the livelihoods of thousands of Kentucky farm families.

The history of tobacco is one of change. The changes described in part 1 are most obviously tangible—the introduction of new technologies, changes in the way tobacco is prepared for market and is sold, and so on. Through my ethnographic research I became aware that within the lifetimes of the farmers whom I came to know, the social and political meanings of tobacco had changed dramatically as well. I became aware of the incongruities between what I was seeing in the media and hearing in conversation and what I was learning from farmers and from others in tobacco communities.[30] Part 2 is a result of my search for how this change occurred.

Chapter 4

Tobacco's Move from Self-Evident to Self-Conscious Tradition

Between the early 1940s, when the Kentucky Department of Agriculture began publishing a newsletter, through the 1960s, the KDA's coverage of tobacco quickly became increasingly self-conscious as threats to tobacco increased. Although the "health scares" began in the early 1950s, they were not directly remarked upon until the release of the surgeon general's report in 1964. Tobacco was occasionally described as a "tradition" in this period, if it was labeled at all. While the "tradition" label marks a practice as self-conscious, it also implies continuity. Tobacco is thus presented as an ongoing practice, even though it is one that is reflected upon.

The Second World War and the Cigarette

The March 1944 issue of the KDA newsletter makes it clear that from the perspective of the Department of Agriculture, tobacco is an embraced commodity. The commissioner of agriculture applauds increased smoking rates in the United States and abroad and urges maximized per-acre production in order to keep up with demand. Both tobacco production and the prices that companies were paying to farmers were high during the Second World War, the period during which smoking rates increased the most in US history,[1] as soldiers smoked cigarettes provided by the armed forces and came home addicted.[2] In December 1944 and January 1945, the outlook for burley tobacco as reported in the KDA newsletter was good, although there was a cigarette shortage in 1944 that merited a congressional investigation. According to the KDA, "just who [was] to blame" for the shortage "[was] difficult to determine," because the government had not released manufacturing and sales data.[3] In 1944, "domestic consumption of tobacco [was] at an all time high," and military consumption was also high.[4] At this time, there was as yet no reason for the KDA to mask its desire to see tobacco consumption continue to rise; rising rates of smoking were simply good for the economic future of the state.

Immediately following the war, prices dropped, and throughout the later 1940s the KDA repeatedly advocated various efforts to increase foreign markets for burley, including the establishment of a new burley marketing organization. Huge crops were grown in 1945, due in part to the increased use of fertilizer and lime, which resulted in increased yields.[5] This led to a drop in prices in 1946, followed by debates over whether or not to lower quotas. This is the cyclical pattern that remained in place through the end of the program: growing demand led to higher prices and calls for farmers to grow more tobacco; farmers grew more and saw prices drop and quotas cut as pool stocks—tobacco not bought and therefore held on loan until it was sold—grew. In January 1946, it was noted that "the recent slump of burley tobacco prices has caused grave concern among all Kentuckians for this crop is more than any other the basis of Kentucky prosperity,"[6] reflecting the importance of tobacco economically not just for farmers but for the general well-being of "all Kentuckians." Throughout this time, through both articles and direct appeals from the commissioner, the KDA actively encouraged farmers to vote to continue the tobacco program in referendums held every three years.

Examples of coverage that treated tobacco as a self-evident aspect of the agricultural economy during this period include coverage of the KDA checking tobacco warehouse scales prior to the start of the tobacco marketing season, information about disease prevention and new farm technologies, and photos of Future Farmers of America (FFA) members selling tobacco. Photos of the young male members of the FFA selling their tobacco reflect the self-evident nature of tobacco at the time, as children's involvement in raising tobacco was not yet questioned. Instead, such photos present the implicit argument that these are the tobacco growers of the future. In April 1946, there was a piece entitled "Profit from Hens Surpasses Tobacco," a story about one farmer who raised a large flock of

KENTUCKY MARKETING BULLETIN

Published by
DIVISION OF MARKETS
DEPARTMENT OF AGRICULTURE
Frankfort, Ky.
ELLIOTT ROBERTSON, Commissioner

Entered as second-class matter October 1, 1940, at the postoffice at Frankfort, Ky., under Act of June 6, 1946

MARCH, 1946

VOL. III, NO. 3

Fourth Annual Marketing Clinic Highly Successful

A Burley Grower's Plan for Promoting Tobacco

One of the most logical plans for promoting Burley tobacco that I have seen or heard was presented over WHAS some time ago by Ivan Jett in an interview with Frank Cooley, Agricultural Coordinator for the station. This plan is worthy of consideration by all farmers and farm groups and should be discussed by them in their county and community meetings.

For the past two years our Burley production has averaged about 620,000,000 pounds annually while consumption is from 425,000,000 to 450,000,000 pounds. The 10% cut in acreage just announced by the Secretary of Agriculture plus the severe penalties for excess production will bring production more in line with present consumption, but wouldn't it be much better if a foreign outlet could be secured for this crop?

Mr. Jett's plan is simple. He proposes

(Continued on Page Four)

Livestock Sales and Shows

Will sell fifty head Registered Herefords, March 30, 1946, at auction. 25 Bred Heifers, 10 Cows, 10 Bulls, ready for service, five open Heifers. Prince Domino, Hartland Mischief, Milky Way and WHR Blood. C. E. Palmore & Son, Bowling Green, Kentucky.

The Cocker Spaniel Club of Kentucky, Inc. will hold its first annual specialty show in Louisville, April 14, 1946. Anyone interested may obtain premium list and entry blanks by writing Mrs. E. Louis Morris, 2203 Edgehill Road, Louisville, Kentucky, show secretary.

Western Kentucky State Teachers College, April 9, 1946, dispersal sale of Purebred Shorthorn Cattle. The sale is at Bowling Green, Kentucky. For further information write Mr. Charles L. Taylor, Acting Head Agriculture Department, Western Kentucky State Teachers College, Bowling Green, Kentucky.

COMMISSIONER'S COLUMN

Shall the O.P.A. Be Abolished?

There are many arguments pro and con on the tenure of the O.P.A. These arguments are not confined to the man in the street, but they are widely distributed in the press, among the radio commutators, business executives, and last but not least, Mr. John A. Q. farmer.

There appears to be much confusion on the question of what the O.P.A. has done and still more confusion on the question

(Continued on Page 5)

Representatives of Distributors, Marketing Experts, and Many Farmers Attend.

High lighting the Fourth Annual Food Marketing Clinic was a talk given by Mr. C. B. Denman, Agricultural Counsel for the National Association of Food Chains. Mr. Denman gave one of the most inspiring and instructive talks ever heard at any of the previous Clinics. Mr. Denman said "Decentralization of processing of products by having more done by the pro-

(Continued on Page 7)

FUTURE FARMERS SELL TOBACCO—Pictured above is a group of Kentucky Future Farmers of the Calhoun-Beech Grove chapter examining tobacco after the buyers moved on at the F. F. A. Tobacco Show and Sale held at Owensboro, Jan. 1 and 7. One hundred and seventy Future Farmers sold 188,058 pounds of dark and burley tobacco at the sale. The dark tobacco averaged $25 per hundred pounds and the burley averaged $40 per hundred. The Daviess County chapter was tops in poundage with 42,950 pounds sold. The Lewisport-Hawesville chapter sold 37,090 pounds, while the Calhoun-Beech Grove chapter sold 30,010 pounds. At the 1945 sale, 106 Future Farmers sold 128,926 pounds of tobacco. W. R. Tabb, executive secretary of the Kentucky Association of Future Farmers of America, is shown at extreme left.

"Future Farmers Sell Tobacco." (*Kentucky Marketing Bulletin*, March 1946, Courtesy of the Kentucky Department for Libraries and Archives)

hens and made a greater profit from them than from his four acres of tobacco. Here and in similar articles in the newsletter sporadically over the years, tobacco income is understood as a yardstick by which to measure other farm income, highlighting its status as the default crop. The July 1948 issue was taken up almost entirely by an article about the inclusion of tobacco in the Marshall Plan for European Recovery in order to secure "the inclusion of adequate quantities of United States grown tobacco in the recovery program."[7]

In July 1948, there was a brief note that tobacco would return to the state fair exhibits after a fifteen-year absence. From this point forward, the rules of tobacco judging and the winners in each class were periodically included in the newsletter, although such details about other exhibits at the state fair were not. This coverage was most consistent from 1948 through 1953; in the years that follow, the coverage was there a year, then gone for three to five, then there for a year, and so on. The tobacco exhibit at the state fair was included in a total of eighteen years (often in multiple issues) out of the sixty-four years of the newsletter's publication.

Another ongoing feature in the newsletter, photos of tobacco auctions in progress, debuted in November 1949. Often these obligatory photos pictured the opening day of sales, and often the commissioner of agriculture was pictured in the line of buyers, auctioneers, and warehousemen during a sale. These photos appeared in nearly half of the newsletter's years of publication (thirty out of sixty-three), with the last one appearing the year before the tobacco buyout, which ended the auction culture almost entirely. The photos served as a visual demonstration of the KDA's role in tobacco sales. The presence of the commissioner and other officials in the photos further argued that tobacco sales (the sale of YOUR tobacco) were the commissioner's priority.

From the beginning of the newsletter through the end of the 1940s, tobacco was written about in an average of over two-thirds of the issues each year, frequently covered on the first page. For instance, it was on page 1 in six out of twelve issues published in 1946. Overall, tobacco coverage in the 1940s is fairly self-evident, even as its role in the agricultural economy was highlighted above other commodities. As the 1950s dawned, however, a more self-consciously celebratory period began as tobacco increasingly came to be understood as threatened.

King Tobacco Reigns

Although the KDA did not mention it directly, the shift to more self-conscious coverage of tobacco in the newsletter coincided with rising attention to medical evidence about the effects of smoking beginning in the early to mid-1950s "that made it abundantly clear that the evidence implicating cigarette smoking as a risk to health was now of a different order."[8] This came from such places as a series of articles published in *Reader's Digest* in the 1950s and '60s, beginning with the 1950 article "How Harmful Are Cigarettes?"[9] and with a report released by the American Cancer Society to the American Medical Association in 1954.[10] Although these antismoking campaigns were not mentioned in the KDA newsletter, they likely motivated an increased self-consciousness about tobacco, as the entire tobacco industry braced for attack.[11] According to Allan M. Brandt, "no major industry had ever faced such a fundamental threat to its future" as the tobacco industry faced beginning in the early 1950s.[12] Rather than attempting to answer critics directly (as it later would), the KDA deployed an economic defense of the industry. This defense placed tobacco squarely in the present, as important to the state *today*.

This began in 1950 through 1951, with a five-part "pictorial serial" following the tobacco growing season, including photos of plant beds and plowing (May 1950); pulling plants and setting them (July 1950); topping and cutting (September 1950); baskets packed at sale and auction (February 1951); and final weighing, packing into hogsheads, and processing at the redrying facility (March 1951).[13] Most of the serial appeared inside the newsletter and did not include textual information about what was happening in the photos, other than a single word or phrase such as "plowing" or "topping" and the location of the tobacco (all on the same farm). However, the final two segments were front-page photo spreads followed by brief descriptions of what is happening in the photos and the role of the KDA in these activities. The auction piece covered the front page with four photos and a large block-lettered headline: "KING TOBACCO REIGNS." This pictorial serial, the first of multiple displays and series like it, marks a shift to a self-conscious presentation of tobacco. No other farm product was treated to such a series in over sixty years of the newsletter.

This serial also provides important clues about the KDA's perceived audience. The series was introduced in May 1950 with a statement that indicates that its direct audience was not imagined as farmers: "We have all seen burley grown but this article is intended to picture the main phase[s] of the growth and marketing of the crop—a graphic record of Kentucky's principal crop."[14] Other articles in

KENTUCKY AGRICULTURAL BULLETIN

Published by
DEPARTMENT OF AGRICULTURE

HARRY F. WALTERS, Commissioner

FRANKFORT, KENTUCKY

Entered as second-class matter October 1, 1940, at the postoffice at Frankfort, Ky., under Act of June 6, 1940.

FEBRUARY, 1951

VOL. VII—No. 2

KING TOBACCO REIGNS

"King Tobacco Reigns." (*Kentucky Agriculture Bulletin*, February 1951, Courtesy of the Kentucky Department for Libraries and Archives)

KDA newsletters (in this and in other issues) speak directly to the farmer through pronoun usage—using *you,* for instance, in articles about technology and farm practices, support for the tobacco program, and so on. Those articles that are intended to teach nonfarmers about tobacco production serve multiple purposes, however, because in addition to educating the nonfarmer, they provide the farmer with a point of identification with the KDA. By seeing the work they do depicted in the pages of the newsletter, farmers would—so the KDA perhaps hoped—see the department (and the commissioner) as their friend and advocate. In this sense, such features as the pictorial serial of tobacco celebrate tobacco work as part of the current way of life of Kentucky agriculture, a way of life that the KDA wished to persuade multiple audiences to understand and preserve.[15] Such serials also claim tobacco knowledge as official knowledge controlled by the KDA.

In 1954, the newsletter's layout changed, and a new masthead was introduced that featured illustrations of farm products, including a large tobacco leaf at the center. While in general all the objects in the masthead are out of proportion to one another (e.g., the ear of corn is as big as the head of the horse), the tobacco leaf is far out of proportion; it is the largest object and is pictured at the center. This masthead, a visual argument that tobacco is central to the farm economy, remained in use until 1969.

In March 1955, the newsletter's front-page headline read, "Tobacco Samples Become Museum Pieces," running below a full-page photo of a "permanent exhibit of samples of all types of tobacco grown in Kentucky in the museum of the Kentucky Historical Society in the Old State Capitol Building."[16] According to the accompanying article, "farmers, school children, tourists and the passerby will now have the opportunity to see for themselves exactly what each grade of tobacco leaves should look like," and "this display will be of particular value to tobacco growers as it will provide them with a guide for sorting their own tobacco."[17] Providing a "guide for sorting their own tobacco" seems like an odd purpose for a museum exhibit, yet this display provides another example of the KDA's claim on what was formerly vernacular knowledge. Much as it had in the pictorial serial, the KDA here was both educating the public about tobacco—and therefore arguing the importance of tobacco—and assuring farmers that their interests were aligned.

According to Barbara Kirshenblatt-Gimblett, "display is an interface that mediates and thereby transforms what is shown into heritage."[18] Tobacco has a long history of display. Most obviously, tobacco was displayed for government graders, buyers, and other farmers at the time of sale. But in addition to functional display for marketing purposes, tobacco has been exhibited for judging at agricultural fairs and festivals. At such events, children and adults displayed their

KENTUCKY DEPARTMENT OF AGRICULTURE

Bulletin

MARCH 1955 VOL. XI. NO. III

See Story on Page 6

TOBACCO SAMPLES BECOME MUSEUM PIECES

"Tobacco Samples Become Museum Pieces." (*Kentucky Department of Agriculture Bulletin*, March 1955, Courtesy of the Kentucky Department for Libraries and Archives) best tobacco—green and on the stick, as well as cured and tied in hands—alongside their best fresh vegetables, cakes and pies, home-canned goods, quilts and other domestic arts, livestock, and other material culture from daily life. In her study of the midwestern county fair, Leslie Prosterman argues that at the county fair, people present their images of themselves and their values through the display of materials from their daily lives—from cattle to pickles. According to Prosterman, "In effect, the county fair recontextualizes everyday life through the exhibition process. It takes work out of one place into another and by this act requests a new kind of notice. That notice in turn grants consciousness to the work within the home, barn, or studio."[19]

Although, as Prosterman argues, such displays bring a consciousness to the fruits of rural labor, this consciousness is different from the self-consciousness that accompanies a move to heritage. County fair displays are intended for members of the community, those assumed to share systems of aesthetics and values, rather than for outsiders. This is most clearly evidenced by a lack of contextualizing information such as the signage and labeling that is obligatory in museum display. Such vernacular displays are a celebration of a bountiful harvest and a show of pride in raising a "pretty" crop of tobacco rather than a display of heritage for outsiders. This 1955 exhibit provides a transition point, as tobacco was put on display for insiders as well as outsiders, making a significant statement of self-consciousness about the importance of tobacco to Kentucky. Placement in a state museum rather than another public site suggests an implicit perception of tobacco "heritage," although this term would not be used in the newsletter for several years to come.

During the mid-1950s, however, the economic rhetoric in defense of tobacco began to intensify as the importance of tobacco not only to growers but to the entire state became a central argument in defense of the crop. In April 1955, the newsletter's front page featured a letter to tobacco growers from Commissioner Ben Adams in which he warned of the "grave danger" that the state would be in if growers did not vote in favor of continuing the support-price program for the next three years. He argued, "No other enterprise in our State has as great an influence on the entire economy of Kentucky as the tobacco industry. Naturally, whenever the well-being of the tobacco industry is in jeopardy it is a matter of concern to everyone."[20] This argument clearly aligned the interests of the state with those of the industry

"Kentucky's Most Profitable Tradition." (*Kentucky Department of Agriculture Bulletin,* September 1955, Courtesy of the Kentucky Department for Libraries and Archives)

and therefore, implicitly, of the major tobacco companies. Although Peter Benson argues that pointing to the economic needs of tobacco communities as a justification for the legality of tobacco "intensified during the 1980s and 1990s,"[21] this argument was in use as early as the 1950s. This economic rhetoric was front and center in September 1955, through a full-page cover photo of a field of tobacco and a tobacco barn with the caption "Kentucky's most profitable tradition," reflecting self-consciousness about tobacco as an ongoing cultural and economic practice.

In May 1956, the assistant agriculture commissioner testified before a US Senate subcommittee against new manufacturing techniques that reduced the amount of tobacco in cigarettes, "including grinding the entire leaf, including stems [which had once been removed], and combining it with a binding or cohesive agent and then rolling the compound into sheets."[22] This practice was obviously bad news for tobacco farmers because it meant companies could purchase less tobacco, but the assistant commissioner also argued that the related decreased emphasis on tobacco quality in product advertising would negatively affect sales.[23]

In his testimony, the assistant commissioner stressed the "dependence of Kentucky agriculture on tobacco" and noted that "soon after John Rolfe began commercial culture of tobacco at Jamestown, Virginia, in 1612, quality became essential."[24] He went on to emphasize the importance of tobacco to everyone, "since tobacco has long been considered a favorable commodity to tax by the Federal and State Governments."[25] The discursive alignment of the historical significance of tobacco with economic benefits to *all* became increasingly prevalent at this time. This reference to history marks the first instance of what soon would become a direct appeal to heritage. The assistant commissioner noted, "Burley tobacco growers have just completed one of the most confusing market seasons in history. After years of efforts to achieve quality desired by smokers, our growers have found that certain lower grades of Burley have been much in demand. Such trends have caused the grade-price relationship to seemingly be overhauled without satisfactory information to make the necessary adjustments along the line of production and marketing."[26] This is one of only a handful of times that company practices were criticized in the pages of the KDA newsletter. More often, company officials appeared in pictures with commissioners and KDA board members, and their campaigns and activities were reported on with praise.

The newsletter's coverage of tobacco dropped as the 1950s progressed; each year, excluding 1954 and 1955, there were articles on tobacco in half or fewer of the twelve issues. The interests and opinions of the state's commissioners of agriculture clearly affected the amount of tobacco coverage, and the drop in coverage in the late 1950s may be traceable at least in part to Commissioner Ben J. Butler, who came into the office in December 1955. With the exception of a photo of him with the previously mentioned tobacco exhibit in the old capitol in March 1958, the slim tobacco coverage was mostly limited to self-evident aspects, such as market reports and instances of warehousemen charged with using faulty scales. In 1959, former lieutenant governor Emerson Beauchamp was introduced as the incoming commissioner of agriculture. His professional experience is listed, but his status as a tobacco man is highlighted, as he is shown in a photo "examin[ing] some bulks of tobacco grown on his farm."[27] The tobacco coverage increased during Beauchamp's term.

KING TOBACCO THREATENED, THE QUEENS REIGN

In December 1960, photos of the opening of the marketing season returned after a four-year absence, with two front-page photos and the caption "Scenes such as the one above are being repeated dozens of times daily as the most important event of the year, the tobacco marketing season, occurs in Kentucky again."[28] This caption marked the opening day as "important" through its many perpetuations, yet normal because it was happening "daily." The February 1961 issue included a photograph of the Kentucky float in the inaugural parade for new president John F. Kennedy, featuring tobacco, thoroughbred horses, and bluegrass. The caption reads:

> Tobacco, Kentucky's most important crop, was combined with some of the State's other fine traditions to carry out a theme of "Peace" for the Commonwealth's float in the Kennedy Inaugural Parade in Washington, January 20. The float displayed a giant peace pipe and some burley tobacco, naturally, to smoke in it. In addition, the float was replete with lovely girls, a Kentucky Colonel and his lady and a replica of My Old Kentucky Home. The horses, of course, represented Kentucky's thorobred [*sic*] industry. The float was carpeted with a facsimile of our Blue Grass and was trimmed with gold foil.[29]

DECEMBER 1960 VOL. XV NO. 12

$38,864 Refunded To Counties By Dog License Section

A total of $38,864.60 has been returned by the Kentucky Department of Agriculture to 73 of the participating counties in the state dog licensing program.

These refunds to the counties represent their share of the proceeds from the sale of 77,034 single tags and 674 kennel licenses during the period July, August and September. Single tags sell for $1.50 annually and kennel licenses for $10-$15 dependent upon the number of dogs in the kennel. Of the $1.50 received from the sale of a license, 25 cents is retained by the county dog warden, 50 cents is refunded to the county where the license is sold, and the remaining 75 cents is deposited into the State Livestock Fund to administer the law and pay indemnities to property owners for damages caused by unidentified dogs.

According to the law, the county shall use its refund for enforcing the law and the operation of a dog pound.

CANNED BEEF & GRAVY ORDERED FOR SCHOOLS

Kentucky children attending schools participating in the National School Lunch Program are getting canned beef and gravy added to their lunchroom menu Commodity Distribution director Tom Lewis stated.

Lewis revealed that 848,250 pounds of beef and gravy has been ordered for delivery this month to the 1,451 Kentucky schools in National School Lunch. The objective of the Program is to provide the children with at least one-third of their daily nutritional requirements.

He stated that this allotment, 15 carloads, is considered as a three month supply on the basis of eight-tenths of a pound per child per month.

Lewis added that the participating schools had already been allocated a four-month supply of frozen hamburger and a two-month supply of frozen turkey. With the addition of this beef and gravy allotment the schools in the lunch program can provide a nutritional diet containing meat for the entire school year.

"Kentucky is receiving a good share of the foods made available through the $60 million National School Lunch Program," Lewis declared, "and it is helping us to serve our children more nutritious lunches."

Scenes such as the one above are being repeated dozens of times daily as the most important event of the year, the tobacco marketing season, occurs in Kentucky again. In the photo above Hogan Teater, wearing light trench coat second from left, "starts" a basket of tobacco at the People's Warehouse in Somerset where he is a part owner. Teater, whose home is in Lancaster, is a second-term member of the State Board of Agriculture.

Another Board of Agriculture member who has an interest in a tobacco warehouse is Mack Walters, Shelby County, part owner of the Globe House in Shelbyville. Walters, on the left, was visited by Commissioner Beauchamp, light coat, on the day that George Fremd, Jr., Henry County, sold part of his 40 acre crop at the Globe House. In the above photo Fremd, who has the largest allotment in his county, gets a basket of his tobacco appraised. Commissioner Beauchamp says that he has a dual interest in the tobacco market. First, as the state official responsible for orderly marketing procedures and, secondly, he is a tobacco grower himself and is well pleased with the good burley market we are having this season.

Expanded Program Of Crop Reports

The Kentucky Crop & Livestock Reporting Service is the official statistical agency for agriculture in the state. This work has been mainly a Federal responsibility, but several individual projects such as the Manufactured Dairy Products Report and Special Marketing Surveys have been carried out cooperatively with the State Department of Agriculture. The year 1960 marks the begin-

(Continued on page 2)

W&M Program Costs 9 Cents Per Inspection

It is costing the Kentucky taxpayers slightly less than nine cents per inspection to safeguard the accuracy of scales, meters and numerical count used in trade, a study of the operation of the Division of Weights & Measures discloses.

Inspection costs range from a low of .9 of a cent for prepackage inspection to a high of $9.30 per unit for checking farm bulk milk tanks. In between these two extremes are clothespins that must be counted; cavair that must be weighed; gasoline to be measured; coal mine scales to be balanced; stockyard and tobacco warehouse weighmen to be bonded; anhydrous ammonia to be safety checked; and, improper or misleading packaging to be prevented.

(Continued on page 2)

Egg Wholesaler Gets Suspended For Violations

The egg wholesaler's license of Clarence Cox, doing business at 7294 Whipple Road, Valley Station, in Jefferson County was suspended for five days for violations of the Kentucky egg marketing law.

Gordon O'Banion, egg law supervisor, said that the suspension, which commenced December 6, would be followed by a 30-day probationary period beginning December 11.

O'Banion disclosed that this action was taken after state inspectors had found that Cox was: 1. Offering eggs not up to consumer grade standards. 2. Selling eggs in cases that were not labelled as to egg grade and size and not bearing the name and address of the wholesaler. 3. Not using the required inspection fee stamp. 4. Not keeping the required records of all transactions.

During the suspension and probationary period Cox will be required to bring his records up to date and secure the required inspection stamps. These stamps are provided by the Kentucky Egg Marketing Section for a fee of two cents per 15 dozen crate.

Failure to comply with these stipulations might result in the revocation of Cox's egg handler's license.

(Continued on page 2)

Opening of the tobacco marketing season in 1960. (*Kentucky Department of Agriculture Bulletin*, December 1960, Courtesy of the Kentucky Department for Libraries and Archives)

Although there was no attempt to hide tobacco's use as a substance that is smoked, the inclusion of a "peace pipe" rather than a cigarette suggests that reminders of the historical uses of tobacco may have been perceived as safer than the increasingly criticized cigarette. *Tradition* continued to be used to describe these products and practices.

Tobacco coverage was prominent in the fall and winter of 1961, with exhibits at the state fair featured in August and December, the tobacco princess in October, and the Carrollton Burley Festival in November; these represent the first time that either the state tobacco princess or a tobacco festival was mentioned in the newsletter. Coverage of tobacco princess and queen pageants had begun in 1960 (with coverage of the dairy princess appearing a few months earlier), with a photo from the International Tobacco Queen Contest in Raleigh, North Carolina, featuring "twenty girls from six tobacco states and three foreign countries," including three Kentuckians.[30] Coverage of tobacco-related beauty pageants swelled over the next decade and then faded. Sporadic coverage of a range of other commodity-inspired titles was present during this period as well.

In July 1962, there was a front-page article about the search for a "Young Lady to Be State Tobacco Princess." Qualifications included being "beautiful in the tradition of Kentucky women."[31] The article went on to note, "This Department is cooperating in this project because Commissioner Beauchamp believes that it provides an opportunity for the promotion of the tobacco industry at a time when the industry is beset by health scares and foreign market complications."[32] This is the first time that "health scares" were directly acknowledged in these pages: at this point the KDA was open about its use of a defensive strategy in the face of the two greatest threats to tobacco in this period: increased attention to the dangers of smoking and overseas production. A defensive posture is also evident in a concession that seems to have been made to critics, whether actual or anticipated: the article notes that although the Department of Agriculture was coordinating the pageant, tobacco organizations such as the Burley Co-op would pay for it, not taxpayers.

Tobacco princess coverage continued to increase in 1963, with articles in April, July, August, and September. By now the state pageant had become a mainstay, with the state tobacco princess going on to compete in a national pageant for the title of "Queen of Tobaccoland." Photos of the young queens and princesses with the

KENTUCKY DEPARTMENT OF AGRICULTURE

Bulletin

AUGUST 1963

VOL. XVIII NO. 8

Dog Loss Checks Mailed to 572 Valid Claimants

A grand total of $36,053.60 has just recently been disbursed to Kentucky property owners for losses due to dogs, Dog Licensing Administrator Karl Vick reports. This money was paid out of the State Livestock Fund to satisfy 592 valid claims presented by property owners in 68 counties. All claims were settled at full market value as determined by the USDA, Vick added.

Vick pointed out that for a claim to be valid the following conditions must be met: 1. The dog or dogs causing the damage must be unknown to the claimant. 2. The county in which the claim occurred must be in compliance with the Dog Law. 3. The claimant must not have owned or harbored unlicensed dogs.

The Livestock Fund is accumulated from the sale of dog licenses. From each $1.50 license sold, 25

(Continued on page 3)

TOBACCO PRINCESS IS A REAL TOPPER

TOBACCO TOPPING is in full swing in Kentucky now but it is doubtful if many farmers have a "topper" like this young lovely in their fields. Miss Janie Olmstead, who is very appropriately the Kentucky Tobacco Princess, "helps out" in the topping on the farm of her grandmother, Mrs. Dorothy Allison, near New Castle. Janie is now a sophomore at the University of Kentucky, but as a small girl she used to pick up tobacco leaves on her grandmother's farm. Janie will relinquish her title during the Kentucky Tobacco Princess Pageant, an opening day feature at the Kentucky State Fair, Sept. 6-14. Governor Combs will be on hand to crown the 1963 Kentucky Tobacco Princess.

Simon Named As Head Of Milk Marketing Agency

Effective August 1, Paul M. Simon became the executive secretary of the Milk Marketing and Antimonopoly Commission. Simon, a seven-year employee of the Department, was selected to fill the vacancy created by the death of W. L. Rouse, on April 7.

Simon, 38, came to the Division of Markets in 1956 as a field

(Continued on page 3)

"Tobacco Princess Is a Real Topper." (*Kentucky Department of Agriculture Bulletin,* August 1963, Courtesy of the Kentucky Department for Libraries and Archives)

commissioner, company representatives, and other older white men became more frequent, as did occasional photos of the women with their prizes (sets of luggage were common). Women had become ambassadors for this historically male-centered industry; their young and healthy bodies were used to combat increasing "scares" about the effects of smoking on the body.[33]

Coverage of tobacco increased yet again and changed in tone upon the release of the surgeon general's report on January 11, 1964. The front page of the newsletter's February 1964 issue included an article celebrating the centennial year of burley, marking the moment described in the origin narrative when white burley was discovered on a Bracken County farm (see the introduction), and excerpts from the commissioner's testimony to the House Tobacco Subcommittee about the need for additional tobacco research. A line along the bottom of the front page reads, "1964—BURLEY TOBACCO'S CENTENNIAL YEAR—1964"; this line would be repeated on at least one page of every newsletter published in 1964, beginning with this February issue.

FEBRUARY 1964 VOL. XIX NO. 2

Butler Talks To State Livestock Group February 20

Commissioner Butler will be one of the featured speakers when the Kentucky Livestock Improvement Association holds its annual meeting here in Frankfort, Thursday, February 20.

W. T. Forsee, Owenton, is president of the association which is the oldest general affiliation of livestock breeders and feeders in Kentucky. The K. L. I. A. has been in existence 28 years.

The Commissioner will speak that morning on the topic: "The State Department of Agriculture's Promotional and Regulatory Programs for Livestock." In presenting this topic Butler will discuss the Kentucky dog licensing law; the livestock disease control program with emphasis on the new hog cholera program; the proposed animal disease diagnostic laboratory; and, the department's plans for the promotion of livestock through its shows and fairs programs.

Another featured speaker on the K. L. I. A.'s annual convention program will be R. B. Carothers, Sr., of Paris, Tennessee. Carothers operates a large ranch in Tennessee and also serves as a director of the National Livestock Feeders Association headquartered in Omaha, Nebraska.

(Continued on page 2)

Shorthorn Show & Sale At Bowling Green March 14

Leading Shorthorn and Polled Shorthorn breeders from throughout Kentucky and adjoining states will compete for more than $9,000 in prize money at the Kentucky National Shorthorn Futurity Show and Sale, March 14, at Bowling Green. The event will be held in the Co-op. Sales Barn located on U. S. 31 W., one mile south of town.

The sale is under the joint sponsorship of the Kentucky Department of Agriculture and the Kentucky Shorthorn Association, Jack Ragsdale, Prospect, president. Each of the groups is offering half of the premium money.

65 Head Offered

Malcolm B. Tucker, Chrisman, Ill., will judge the cattle at 9 a.m. March 14. Thirty-two leading

(Continued on pages 2 & 3)

Department Participates In Fairs and Horse Shows Convention

Barren County Fair manager, James Owensby (left), Glasgow, receives a trophy from Commissioner Butler for "the most improved fair in 1963." The award was made at the annual banquet of the Kentucky Association of Fairs and Horse Shows in the Kentucky Hotel at Louisville last month. Owensby also is president of the Barren County Fair Board.

Centennial Year Of Burley Tobacco

This spring marks 100 years since a discovery on a farm near the Ohio village of Higginsport, directly across the Ohio River from Bracken County, Ky., set the stage for the chain of events which led to the sprawling, multi-million dollar burley tobacco industry of today.

The year 1964 has been recognized as the "Centennial Year of Burley" by the board of directors of the Burley Tobacco Growers Cooperative Association, which adopted the first resolution "to officially recognize and urge the official and universal recognition" of the year as the centennial.

Called White Burley

At its "birth" today's burley was named "white burley" and its beginning was simple. Practically without exception the strains now grown, of which there are many, come within the generic term "White Burley."

The circumstances of who, precisely, discovered "white burley" seem to be a subject for some discussion in Higginsport even today, with the names of George Webb and Joseph Fore most prominent, but it is the discovery and eventual development and spread of the burley that is monumental.

Came From Ky.

It seems to be without question that the seed from which the "new white burley" grew came from Kentucky -- obtained from George Barkley of Bracken County.

One account written by a man of the time has Webb obtaining the seed and discovering the new, odd plants, while another has Fore doing the job, and still others credit both.

The following account of white burley's "birth" and development was written in 1875 by A. F. Ellis, who was a neighbor of Webb at the time, and it appears to be the best available information on the discovery.

Mr. Ellis wrote: (in part)

"White burley tobacco first made its appearance in the year 1864, near the village of Higginsport, Brown County, Ohio. In the spring of that year one George Webb procured from G. W.

(Continued on page 2)

Hog Cholera Seminar At Bowling Green March 4

Livestock Sanitation officials from most of Kentucky's seven border states are expected to attend the Hog Cholera Seminar that will be held at Bowling Green on March 4, Dr. R. J. Henshaw announces. The all-day seminar will be conducted at the Western Hills Motel in Bowling Green.

According to Dr. Henshaw, the objective of the seminar will be to better acquaint the state's livestock industry with the Kentucky hog cholera program that is scheduled to get underway July 1. Responsible officials from other states will also be familiarized with the Kentucky program.

Henshaw said that in addition to the staff of the Division of Livestock Sanitation, other Kentuckians in attendance would include practitioners, Agricultural Extension personnel and swine producers.

(Continued on page 2)

Butler Testifies To Need For Tobacco Studies

(Excerpts of Commissioner Butler's remarks to the House Tobacco Subcommittee in Washington, Wednesday, Jan. 29. The purpose of this congressional hearing was to determine the need for additional research in tobacco.)

Mr. Abbitt and Committeemen:

My name is Wendell P. Butler, and I am Commissioner of the Kentucky Department of Agriculture. My interest in this subject is due to the responsibilities that our agency of state government has for preserving and promoting the welfare of Kentucky agriculture and the impact of the tobacco industry on the economy and the general well-being of our state and nation.

On January 13, I asked for a federal crash program of research to evaluate the voluminous private studies financed by the tobacco companies and health agencies. I also suggested that the new National Tobacco Research Laboratory located at Lexington, Kentucky, would make a most adequate facility for an exhaustive research project of this type--at a minimum of cost to the federal government.

The tobacco business is at a

(Continued on page 2)

1964----BURLEY TOBACCO's CENTENNIAL YEAR----1964

"Centennial Year of Burley Tobacco." (*Kentucky Department of Agriculture Bulletin,* February 1964, Courtesy of the Kentucky Department for Libraries and Archives)

This front-page attention to tobacco suggests multiple rhetorical responses to the surgeon general's report. The first response, a reminder of tobacco's history, is indirect but can be read as a defense nonetheless, particularly since no such celebration appeared in the January 1964 issue. According to the commissioner of agriculture's testimony, "The tobacco business is at a crisis."[34] With this statement came the explicit crisis-laden rhetoric of *tobacco under attack* that would continue into the 1980s. The commissioner also offered a mantra of defense listing all the things that tobacco had paid for: "In my state, tobacco prices set real estate values; bring good times, or bad, for our economy; furnish additional money to go to Detroit for autos, trucks and tractors; permit homes to be modernized; and often times make a college education possible for deserving young men and women."[35] This mantra is ubiquitous today (a point to which I will return in part 3).

Most important in this period, perhaps, was the commissioner's call for more tobacco research, a call that springs right out of the tobacco industry playbook. Allan M. Brandt demonstrates that the tobacco industry successfully engineered a "controversy" about the effects of smoking beginning in the early 1950s, a strategy that proved remarkably successful for decades: "So long as there appeared to be doubt, so long as the industry could assert 'not proven,' smokers would have a crucial rationale to continue, and new smokers would have a rationale to begin."[36] A significant way this "controversy" was fostered was not only through the industry's flat-out denial of the dangers of smoking but through its ongoing calls for new research in order to argue "that existing studies were inadequate or flawed."[37] The tobacco companies came together in 1953 to establish the Tobacco Industry Research Committee (TIRC), which claimed to conduct research into tobacco's safety, when in fact "most sponsored products had nothing to do with smoking" and "in its first year of operation the TIRC budget approached $1 million—almost all of which went to" advertising and administration.[38] By 1960 "the tobacco industry had succeeded in creating a 'cigarette controversy' within the American media."[39] The KDA's deployment of the industry's rhetoric of calling for "more research" would continue in the pages of the newsletter over the following decades.

In March, the Kentucky commissioner of agriculture spoke directly back to the surgeon general's report, warning of the possible economic crisis that might result from "a campaign against tobacco" and pointing out "the magnitude of the anti-tobacco attacks now under way." "We must vigorously defend our tobacco industry against these attacks," the commissioner said.[40] He noted that the surgeon general had recommended further study in his report, and he expressed agreement—as well as tobacco industry rhetoric—by stating, "It is to the interest of all concerned that no verdict be returned until the question has been resolved by scientific means."[41] This same issue of the newsletter included an announcement that the KDA would once again be coordinating a tobacco princess pageant. The contest was featured on the front page of the June and October issues, as well as in the August issue.

Tobacco princesses and queens again reigned in 1965, with articles about local and national pageants in five issues. In 1967 the tobacco pageant reached a new height, as the Kentucky tobacco princess became the "Burley Belle," and a four-day festival and conference was planned around the August pageant. This festival was announced in March, the KDA noting that the decision had been made to change the pageant's name because sponsorship of the event had long been in the hands of burley organizations, and thus "the annual activity [should] be identified specifically as a burley promotion activity."[42] This name change also served as a reminder of the symbolic importance of burley over other types of tobacco at the state level. The schedule of events was announced on the newsletter's front page in May 1967, along with plans to televise the final portion of the pageant. In July, the newsletter announced that congressmen from other burley states also planned to attend the conference and festival. The pairing of the Burley Belle pageant with this tobacco conference confirms that it was no accident that the height of the burley pageant era coincided with the period of the most direct KDA responses to critics of tobacco's impact on health.

In August, the newly crowned Kentucky Burley Belle was pictured with the commissioner of agriculture on the newsletter's front page, along with her biography and a summary of the pageant. Another front-page article led with the headline "Conference Airs Problems Facing Burley Industry," then went on to characterize the major topic of discussion at the conference attached to the burley festival as "the so-called 'health problem' and the attack on the industry resulting from the surgeon-general's report calling tobacco injurious to health."[43] Kentucky governor Ned Breathitt was quoted as saying, in response to an apparent call for a ban on smoking, that just as taking away food is not the solution to obesity, prohibition is not the answer to questions about smoking and health. With this statement the governor deployed another industry strategy, "flattening what is unique about tobacco in comparison to other [harmful] products," such as fast food, alcohol, and guns—not just any products, "but a subset of products where risk and harm are the subjects of public debate."[44] According to Governor Breathitt, "It is obvious that many people desire tobacco and get satisfaction from the use of it. It is up to research to help find the answers so that this centuries-old custom can continue to please those who want to use tobacco."[45] Here and occasionally elsewhere, not only was tobacco farming defended as a tradition, but so was smoking. "Research" was called upon to validate and perhaps recover both traditions.

In October 1967, the Burley Belle was once again visually positioned in the newsletter next to a report on the conference and discussions about the "constant attacks the industry is undergoing from health organizations."[46] Although there was no mention of a conference in 1968, there were articles about the Burley Festival and pageant in six issues, and the Burley Belle was pictured on Kentucky's float in the inaugural parade in Washington, DC, where she "rode on the Kentucky float along with other representatives of Kentucky tradition not the least of which was a replica of My Old Kentucky Home."[47]

Throughout the 1960s, farming practices such as tobacco stripping techniques, variety selection, and new technologies continued to appear as self-evident aspects of tobacco production as well as further examples of the KDA's efforts to claim authority over tobacco knowledge. In the 1960s, individual farmers were raising more and more tobacco, and barn space was becoming a problem, despite the fact that overall production had begun to decline by this point. The KDA's regulatory role received consistent coverage—the testing of warehouse scales, dates and guidelines for marketing season openings, daily market reports during the sales season—along with policy issues such as the triennial referendum over whether to continue the program. But the most telling trend of the 1960s was the swell of tobacco coverage beginning after release of the surgeon general's report. The KDA was not hiding from tobacco's critics; that would come later. In this period the KDA responded to threats against tobacco by talking about it more, not less. Tobacco coverage increased, from about half the issues each year to every issue in 1967; tobacco appeared on the front page of the newsletter increasingly throughout the 1960s. Photos of the opening day of tobacco markets, which had appeared in only three years during the 1950s, appeared in every year of this decade except 1963, the year before the report was issued. Discourses I have labeled as implicitly self-conscious and heritage-based increased over the course of the decade. Not until the 1970s, however, did overt deployment of the rhetoric of heritage begin.

Chapter 5

Tobacco under Attack

Hello, "Heritage"

A 2003 editorial lamented, "They tell us change is good, and we generally don't shy from it. But when change causes the demise of a tradition, it's a sad thing. Tobacco is not the most politically correct crop in the world, but it is such a huge part of our history and heritage."[1] Here tobacco is described as a dead tradition, killed off by change. By the next sentence the stigma now associated with the crop is referenced, and tobacco is relabeled "heritage," calling forth Barbara Kirshenblatt-Gimblett's definition of heritage as "the transvaluation of the obsolete, the mistaken, the outmoded, the dead, and the defunct."[2] As the rhetoric of the Kentucky Department of Agriculture continued to shift through the 1960s and into the 1970s, the deployment of "heritage" as a defense of tobacco production became pronounced. However, heritage as a defense not only failed to recover tobacco production in the present but also rhetorically recategorized it as a past practice even as it continued.

Tobacco coverage had become increasingly self-conscious by the 1960s, as the "attacks" on the industry increased and the KDA directly responded to the surgeon general's report on smoking and questions regarding health. In contrast, by the late 1970s, the topic of smoking and health was ever present but not always directly acknowledged, although the argument for more research on the health effects of smoking continued to be made. Direct references to the surgeon general's report dwindled from few to none, and articles about tobacco pageants and state fair tobacco exhibits disappeared. Attention turned from health to "attacks" from more generalized "anti-tobacco" forces, as well as calls for increased tobacco excise taxes. However, articles that expressed pride in tobacco were on the rise, and *heritage* entered the lexicon—not just heritage implied but the term itself—joining the economic rhetoric. Heritage and economics would later become discursively separated, and eventually the use of heritage in defense of tobacco would disappear from the pages of the KDA newsletter. By that time, however, "heritage" had become the primary screen through which tobacco was viewed.

Hello, "Heritage"

The 1970s began with consistent tobacco coverage resembling that of the 1960s. The December 1970 issue included a full-page spread on a tobacco harvester that was being tested, an indication of a farm-labor shortage. The cover photo of the 1969 *Kentucky Agricultural Statistics and Annual Report* (advertised in the KDA newsletter in August 1970) depicted a field of tobacco, emphasizing the central place of the crop in Kentucky agriculture. In October 1970, the KDA announced that "talent will not be a factor" in the 1970 Burley Belle pageant. Instead, the contestants would be judged on their recitations of the essay each was to write on the topic "What Burley Tobacco Means to Kentucky," a competition "designed as a means of encouraging general interest in the importance of burley tobacco as the major cash crop in Kentucky."[3] It seems that the healthy young bodies of the Burley Belles were not enough; these "goodwill ambassador[s] for the industry" were now charged with addressing their implicit task explicitly. This attempt was to be short-lived, however. Just a couple of years later the pageants would disappear from the pages of the newsletter. In November 1972, a photo of a KDA staff member who represented Kentucky in the Queen of Tobaccoland contest and a piece on "Miss Tri-State Tobacco," crowned at the Tri-State Tobacco Festival (Ohio, West Virginia, and Kentucky), were the last newsletter features to mention tobacco pageants, although other pageants, such as those for dairy princesses and pork queens, continued into the 1980s. Importantly, tobacco pageants did not end at this time; only coverage of them in the KDA newsletters ended.

In November and December 1970 a two-part series focused on the early history of tobacco in Kentucky at Boonesborough, settled in the late eighteenth century. This two-part series, entitled "The Birthplace of Kentucky's Largest Industry," seems to have been another way of answering the question posed to the Burley Belle contestants, and together coverage of the Belles and Boonesborough represent the beginning of a move from direct responses to tobacco health critics to the deployment of the celebration of heritage as a line of defense. The publication of historical narratives such as this one was on the rise in this period more generally, with numerous examples including a booklet entitled *The American Tobacco Story,* published by the American Tobacco Company in 1960, and a "Tobacco History Series," consisting of short volumes about each tobacco-producing state, published by the Tobacco Institute in the 1960s and 1970s.[4] A number of the book-length histories discussed in the introduction to this volume were also published during this period.[5] The KDA's attention to tobacco history should be read as one among many examples of the deployment of tobacco heritage as a defense against threats to the industry.

Policy issues continued to be discussed, such as a 1971 grower referendum about the move from acreage to poundage, an attempt to address the oversupply of burley resulting in part from the increase in per-acre yields. By moving to a poundage system, the argument went, the program could more closely control the amount of tobacco produced. Earlier attempts to change the burley tobacco program from allotments based on acreage to allotments based on poundage had taken place in the1960s but failed. This time, growers were to be given the choice to vote for the change or to lose the program altogether; the referendum passed.

The February 1971 issue announced the formation of the Council for Burley Tobacco to deal with marketing issues posed by the continuing period of oversupply, with the explicit goal of saving the tobacco program. In August 1974, however, the KDA reported on a "burley heritage print" issued by the Council for Burley Tobacco, marking the first instance I was able to document in the newsletter of the use of the word *heritage* prominently linked to tobacco. The print was called "Burley Tobacco—the Golden Leaf," and it was intended as "a symbolic representation, in the art form, of the burley farming industry's heritage and importance as an economic generator."[6] This depiction of a tobacco field and barn was "developed because there was relatively little art available that was symbolic of the heritage of the burley tobacco industry and its far-reaching effects." Although the council was formed to deal with issues of overproduction, its role quickly became one of marketing burley to both the companies and

Burley heritage print issued

"Burley Tobacco - The Golden Leaf," a limited-edition print of a painting by Kentucky Heritage Artist John Stamper, has been issued by the Council for Burley Tobacco as a symbolic representation, in the art form, of the burley farming industry's heritage and its importance as an economic generator.

The color print embodies the familiar, and nostalgic, scene of growing burley and barn in the background.

Only 1,000 copies of the print are being made available by the Council which developed the project as a long-lasting symbol of the major farm industry. Signed and numbered prints, totaling 500, will be sold for $15.75, including sales tax, and signed but unnumbered prints – the remaining 500 – will sell for $10.50.

The print measures 18 by 24 inches.

The Council's "print project" – basically a promotion idea – was developed because there was relatively little art available that was symbolic of the heritage of the burley tobacco industry and its far-reaching effects.

The edition was "limited" to give it preservation value.

A "miniature" copy of the print, measuring five by six inches, will be distributed without charge as long as supplies last. Persons interested in a small copy of the print may send a postcard to the Council for Burley Tobacco, P. O. Box 2059, Lexington, Ky. 40501.

The Council for Burley Tobacco is a "unity" organization of the various groups and organizations concerned with burley tobacco, its production, sales and marketing, and its preservation as a major farm industry in the areas where it is produced. Kentucky is the major producer of burley.

The Council operates various burley promotion programs, including radio broadcast commercials, print-media advertising, and other efforts to reflect to the general American public the value of burley tobacco production and its economic impact, thereby enhancing the respect for the farmers who produce it and their contributions to the economy.

The "print project" is simply another effort to heighten the respect for the industry and its heritage, said Jack Lewyn, secretary-treasurer of the Council.

THE GOLDEN LEAF—Commissioner of Agriculture Wendell Butler and two staff members, Harold Harberson, left, tobacco specialist, and Philip Smith, director of Markets, look at a print of "Burley Tobacco-The Golden Leaf," painted by John Stamper, Lexington. The painting was commissioned by the Council for Burley Tobacco to recognize the role which burley tobacco has played in Kentucky's agricultural heritage. A limited number of prints are available from the Council's office in Lexington.

AGRICULTURAL NEWS, August, 1974

"Burley Heritage Print Issued." (*Kentucky Agricultural News,* August 1974, Courtesy of the Kentucky Department for Libraries and Archives)

the general public, "enhancing respect for the farmers who produce [burley] and their contributions to the economy."[7] Here, not only do heritage and economic relevance become tied together, but the perceived need to "enhance respect" for tobacco farmers provides an example of a strategy that would continue: calling attention to the farmer in order to align the industry, now threatened, with the wholesome, empathy-inducing image of the farmer.

This print symbolizes a shift from displays of tobacco leaves, as described earlier, to visual representations of tobacco work and of the historical narrative of tobacco. This shift from displaying the thing itself to displaying symbols of it demonstrates increased self-consciousness about tobacco. Paintings and drawings of tobacco landscapes as well as tobacco work—burning tobacco beds or hanging tobacco in the barn, even serial depictions of the entire crop year from seed bed to housing to auction—have become ubiquitous in homes and offices today.[8] The introduction of these prints can also be understood as a rhetorical shift from "land" to "landscape," for as Gregory Clark argues, "*land* becomes *landscape* when it is assigned the role of symbol, and as symbol it functions rhetorically."[9] The current nostalgia for the fading tobacco landscape—disappearing tobacco fields and collapsing tobacco barns—can be understood as commemorating a *heritage landscape* and mourning its perceived passing, part of the rhetorical discourse that argues that tobacco is going.

Another historical moment was featured in November 1972, the fiftieth anniversary of the founding of the Burley Co-op, in an article submitted by the co-op about its history of working to "uplift the lives of the growers of burley tobacco."[10] The following month the standard report on the tobacco auction market was expanded to include another pictorial description of the process from barn to warehouse, headlined "Burley Payday." In addition to a report on burley sales, the accompanying text describes the tobacco work taking place in the photographs for the nongrower audience. Although coverage of tobacco dropped in 1974—seven issues of the newsletter contained no tobacco articles; tobacco did not appear on the front page once; and the newsletter featured no obligatory articles on the opening of the markets, scales being checked, the burley outlook, or state fair tobacco exhibits—tobacco coverage increased again in 1975, with only two issues lacking attention to it. Much of this coverage can be categorized as self-evident, such as reports on quotas and the auction system and new tobacco baling machines being tested by the University of Kentucky.[11] An article about a man who continued to practice the "disappearing art of hand-splitting tobacco sticks" suggests heritage through its focus on a vanishing aspect of tobacco history.[12] Tobacco politics was covered, including the then-current investigation into "alleged improprieties of the current [auction] system" by a legislative subcommittee that had agreed that "a system of allocation has replaced the traditional auction . . . [but] found no direct evidence of collusion among the buyers or tobacco companies."[13] This issue came up increasingly in the last decades of the tobacco program, and today it is common to hear farmers remark that by the end the auction system was no longer a "true" auction.

8—KENTUCKY AGRICULTURAL NEWS, December, 1972

Burley Payday

Burley sales recessed December 14 for Christmas, and farmers who have not sold their crop hope the layoff might bring higher prices.

It is estimated that some 75 percent of the 1972 crop is already sold. Most farmers, some observers say 90 percent, have gone over their quota this year. Most of the farmers who were short last year have been able to make up that shortage this year.

Prices during the pre-Christmas selling period were not as high as most farmers expected. High moisture content, mixed grades, fewer foreign buyers — these are some of the reasons blamed for lower than expected prices. Other observers say that record prices from last year led farmers to expect even better prices this year.

Effective quota for Kentucky growers this year is 396 million pounds. More important than the total poundage, of course, is the money they represent — generally considered to be around the 300 million dollar mark annually.

To earn these dollars, a farmer must spend many hours of labor. To capture the final steps of burley production, the News photographer visited with Ira Best, Route 7, Franklin County, as he took his tobacco to market. (Ira's wife Ella, is an employee of the Department of Agriculture.)

In the upper photos, Best, in barn, hands a stick of tobacco to Howard Harley, Shelbyville. In left, middle photo, the truck is being unloaded at the warehouse. In right, middle photo, the tobacco is removed from the sticks, and placed onto the baskets. Once a basket is completed — it cannot be over five feet in height, nor over 700 pounds in weight — is is moved on a "duckbill" to the scales.

After each basket is weighed, it is placed into rows to await the sale. Prior to the sale, it is graded by a USDA grader and recorded with a support price. The next step is not shown, but that is the auction itself, led by the sales manager, followed by a loud chanting auctioneer, who in turn is followed by buyers, farmers and interested spectators. All of this — proceeded by 12 months of labor — must take place before the grower can finally step up to the pay window and experience another BURLEY PAYDAY.

"Burley Payday." (*Kentucky Agricultural News,* December 1972, Courtesy of the Kentucky Department for Libraries and Archives)

Two economic threats began to receive increasing attention in the mid-1970s: taxes on tobacco products and overseas tobacco production. In January 1976, the Kentucky commissioner of agriculture aligned the state with farmers in a fight with the industry that would long continue: he demanded an accurate accounting of foreign imports based on the types of tobacco American companies purchased. He argued, "Here in Kentucky, burley tobacco is, as we all know, the economic mainstay of our farmers"; "Growers want a tariff on imported burley that will protect their interests."[14] In April, Senator Wendell H. Ford—who became known as "Mr. Tobacco" because of his advocacy efforts from 1974 through 1999—testified against a proposal to increase the federal cigarette tax to fifty cents a pack, focusing his testimony on the economic importance of tobacco. He deployed the industry strategy of questioning the legitimacy of existing scientific research, advocating against taxing products that *might* be harmful, and arguing that taxing tobacco was really an underhanded attempt to "kill the tobacco industry."[15] The same issue contained an opinion piece by a representative of the Council for Burley Tobacco, who pointed out that "a dollar received for tobacco, and other farm products, multiplies five times in economic patterns before it fades away" and that "since earliest settlement, tobacco has been an economic mainstay" in the eight-state burley belt.[16] The theme of the multiplication of the tobacco dollar in the local economy became recurrent at this point, not only stressing tobacco's importance to the economy but rhetorically tying tobacco and community together.

Coverage of the tobacco exhibit at the state fair returned in September 1976. A "model of the Liberty Bell made entirely of tobacco,"[17] a fitting representation of tobacco in the US bicentennial year, was also on display at the fair. Although the article does not mention the origin of the bell, R. J. Reynolds had presented a "250-pound replica of the Liberty Bell constructed of pressed tobacco leaves" to the Smithsonian in celebration of the American bicentennial;[18] perhaps this state fair replica also came from Reynolds.

Although he did not mention the bicentennial directly, the commissioner deployed the economic heritage argument in his column in November 1976, echoing the historical narrative promoted in industry publications. He wrote: "Tobacco has had a powerful influence, since the earliest days of Kentucky's settlement, in shaping the economic and social life of our Bluegrass State. The realization of tobacco as a salable agricultural product was an important factor in permanently locating settlers in our rich soil, in building new roads, and in establishing new towns and centers of economic activity. Its potential as a commercial crop [led] to the 'new West's' first export trade and the introduction of Kentucky to all continents of the world."[19] The economic defense is explicitly tied to "social life" and history, marking a further move to self-consciousness about not only tobacco's economic importance but tobacco as responsible for a way of life. He discussed the amount of money that tobacco was expected to generate that year, gave the "multiplies five times" scenario, and concluded, "Whether or not we as individuals use tobacco products, we all live better in Kentucky because of tobacco."[20] This last argument came into increased usage at this time—the idea that some people choose to use tobacco products and some do not; that there is a freedom to smoke; and that despite personal choices, the economic benefits of tobacco to the state are enjoyed by all.

The rhetoric of smoking as a "choice" had been a major tactic of the tobacco industry for decades by this point. In the 1930s and 1940s the "engineering of consent" was coined by an American Tobacco Company public relations consultant working to market Lucky Strikes to women "as the symbol of the independent feminist and the bold, glamorous flapper."[21] This rhetoric had intensified by the 1970s, as "the companies successfully utilized a deeply held traditional American cultural norm that held individuals responsible for their health."[22] By the 1970s, in the context of discussions of the need for increased government regulation of tobacco products, this translated into arguments that "consenting adults" were fully aware of the risks and should be allowed to make their own choices.[23] This argument ultimately laid the groundwork for the industry's successful defense of itself in the lawsuits brought against it by smokers in the 1980s.

By 1977, a new threat arose: accusations that tobacco was subsidized by taxpayer money, in a widespread perception of the tobacco program as a subsidy. In his column, the commissioner of agriculture wrote:

> The tobacco industry has been under attack for some time, but recently a lot of misinformation has been put out. I don't know if this was done accidently or deliberately, but in either case the record needs to be set straight.

> Recently there were reports that the Federal Government subsidized the Burley market. THIS IS NOT TRUE. There have been reports that the Burley program was costing the Federal Government and the taxpayers money. THIS IS NOT TRUE.[24]

He goes on to list the funds borrowed from and paid back to the Commodity Credit Corporation (when tobacco was not sold above the support price and was therefore bought at the support price by the pool) between 1940 and 1975. He also lists the interest paid on the loans and the annual revenue of the federal government through excise taxes, and he states, "so not only does tobacco not cost the taxpayers of this nation any money, it actually makes money."[25] Eventually (in 1982), federal "no-net-cost" legislation shifted the costs of the program to fees paid by farmers and buyers, although the debate about how tax dollars subsidized the industry remained through the end of the program.[26]

In 1978, the rhetoric of the commissioner and others became increasingly aggressive, focused on the "attacks" on tobacco and the need to "fight" the anti-tobacco forces. R. J. Reynolds's "Pride in Tobacco" campaign was the focus of both the commissioner's message and a separate article in the November 1978 newsletter, providing yet another example of the alignment of the KDA with one of the largest tobacco companies.[27] The commissioner explicitly paired heritage and economy: "Let's not lose sight of just how important this great crop and the people who grow, buy and process it really are to the heritage and economy of our commonwealth."[28] He cited the "individual freedom" to smoke and the "harassment" that smokers faced and concluded, "I call upon all Kentuckians who farm, sell, transport or process our crop to fight to retain tobacco as a viable product of our agriculture community. Kentucky needs tobacco and tobacco needs the support of all Kentuckians."[29] Farmers were explicitly charged with the responsibility of fighting for the crop and the industry. The R. J. Reynolds "Pride in Tobacco" campaign was described in another article as "an information program designed to unite the tobacco community to support the region's most important agricultural commodity," "prompted by increasing anti-tobacco pressure" and "critics [who] have been free with words and loose with facts."[30] Materials "outlin[ing] the tobacco industry's side of the public smoking situation, tobacco's impact on the economy, current

Information program gets underway

Tobacco critics 'loose with facts' claims 'Pride in Tobacco' chairman

"Pride in Tobacco," an information program designed to unite the tobacco community to support the region's most important agricultural commodity, was expanded to Kentucky and six other tobacco states Nov. 1, by the R. J. Reynolds Tobacco Company.

William D. Hobbs displays the logo for the "Pride in Tobacco" campaign. The thumbs-up sign and tobacco leaf background are symbolic of the hope for the tobacco industry.

William D. Hobbs, chairman of the North Carolina-based company, told a gathering of press, state officials and tobacco industry leaders that the best way to preserve Kentucky's tobacco economy is for those who rely on tobacco to take an active role in supporting the industry.

"Pride in Tobacco" is an information program geared to the agribusiness community," Hobbs said. "We hope, through this ongoing effort, to make everyone more aware of the importance of tobacco—to growers, manufacturers, related industries and to the entire economic and social fabric."

The announcement of the program was simultaneously transmitted live to five cities in Kentucky. A special telecommunications system was used to link Frankfort with Owensboro, Carrollton, Glasgow, Lexington, and Louisville.

Deputy Commissioner of Agriculture John Anderson welcomed the program on behalf of Commissioner Tom Harris, and urged all members of the Kentucky agribusiness community to give their full support to implementing the program.

Anderson emphasized that tobacco accounts for over 46% of the cash crop receipts in Kentucky, and that it is a significant agricultural commodity in 118 of the state's 120 counties. He added that tobacco is still largely a family farm operation, and that many of these farms would cease to exist without the tobacco industry.

Hobbs said the "Pride in Tobacco" program was prompted by increasing anti-tobacco pressure.

"Many of our critics have been very free with words and loose with facts," he added.

It is expected that every tobacco warehouse will help distribute informational brochures and other materials which Hobbs said will outline the tobacco industry's side of the public smoking situation, tobacco's impact on the economy, current smoking and health facts, and the tax burden on tobacco.

"Tobacco Critics 'Loose with Facts' Claims 'Pride in Tobacco' Chairman." (*Kentucky Agricultural News*, November 1978, Courtesy of the Kentucky Department for Libraries and Archives)

smoking and health facts, and the tax burden on tobacco"[31] were to be distributed through tobacco warehouses. The symbol of this campaign—a cartoonlike hand in the "thumbs up" position, a tobacco leaf, and the phrase "Pride in Tobacco"—was printed on banners, posters, cigarette lighters, and other items. *Pride* is a term that farmers use to refer to a mastery of cultural practices. Through its use of the term, R. J. Reynolds reminded farmers of the need for quality tobacco despite changing cultural practices and growing threats to the industry. The deputy commissioner of agriculture "added that tobacco is still largely a family farm operation, and that many of these farms would cease to exist without the tobacco industry." Here the KDA spoke to tobacco farmers and nonfarmers alike, rhetorically aligning farmers with the industry in their defense of it. Increasingly the tobacco industry—with the support of the KDA—attempted to shift the attention of tobacco's critics from the industry to farmers. Through attempts to put an empathy-inducing face on tobacco, farmers became implicated in the debate.

Diversifying the Farm Economy

By 1979 another shift was beginning to become evident. Although the pages of the KDA newsletter continued to remain an arena where tobacco was defended, it was increasingly defended by individuals outside of the KDA itself, and the agency began to focus more on diversifying away from tobacco. There had been coverage of alternatives to tobacco over the decades, but it was not until the 1980s and 1990s that the KDA appeared to embrace diversification as a true goal for the state's agricultural economy. This coincides with the industry's increasing loss of credibility in the 1980s and 1990s, pointing to a strategic realignment by the state.[32]

In July 1979, the newsletter's cover story pertained to a new Kentucky agricultural products label developed by the Governor's Council on Agriculture. The article began, "Most people associate the label 'Kentucky' with a wide variety of commodities, including Thoroughbred horses, burley tobacco and bourbon whiskey. Now, the state's sellers and manufacturers of agricultural products have a 'Kentucky Connection,' a connection linking their products with a distinctive, standardized market."[33] This seal was to be used on food products made in Kentucky such as honey, eggs, and country ham—not, so the implication goes, on tobacco, horses, or bourbon. This new made-in-Kentucky seal suggests an attempt to create new agricultural associations with Kentucky, and it was the only time that new associations were explicitly contrasted with tobacco, bourbon, and horses. As ongoing attempts to create new associations were launched—attempts that continue today through the KDA's Kentucky Proud program—the contrasts were insinuated but not remarked upon.

In 1979 and 1980, the articles that had come to be standard appeared: opening day of the markets, farming practices and technology, the burley outlook (the pool had no reserves for the first time ever, which should make for a good year for farmers), and a guest column from the Burley Co-op that reminded tobacco growers to vote for the program in the upcoming referendum. The co-op argued that the vote was particularly important for several reasons, including the claim that "anti-tobacco forces in each of the last three years have stepped up legislative efforts to end government loans for tobacco supports."[34] In the past, the commissioner of agriculture would have made this argument in his column. However, Commissioner Alben Barkley, who had come into office in 1980, had a markedly different tone from his predecessor—he did not deploy the "attack" rhetoric as much and did not issue opinions on issues as often. Instead, he left these arguments for guest columnists. For instance, in August 1981 the commissioner described the then-current debates about how best to package tobacco. As described in chapter 2, the move from tying tobacco in hands to packaging it in bales marked an important and controversial moment in tobacco history. Yet in notable contrast to how such issues had been handled by his predecessors, Barkley provided no firm opinion on the shift, concluding that "the system I want is the system that is best for the tobacco farmer."[35]

In the early 1980s, tobacco coverage was primarily limited to obligatory articles, and increasingly, the pieces that continued the passionate defense of tobacco were offered not by the KDA directly, but by guest columnists such as Senator Wendell H. Ford or representatives of farm organizations. This suggests the beginning of a political realignment for the KDA, as well as for individual commissioners—individuals who, after all, serve in elected positions and must therefore consider public opinion. And public opinions about tobacco were clearly changing during this period.

According to Peter Benson, "Honesty about smoking risks became politically expedient in the 1980s."[36] In July 1984 the KDA newsletter made a move from questioning the research to acknowledging evidence that tobacco use is harmful. Tobacco was "becoming a less profitable crop," and "there seems to be 'a substantial block of evidence that smoking is harmful to some people' the Commissioner said."[37] This new commissioner, David E. Boswell, encouraged alternative crops "for those farmers who no longer find it profitable to grow tobacco"[38] and assured readers that the KDA would be working to guarantee markets for these new crops. Alternative crops joined tobacco as the focus of the KDA until the late 1990s, when new crops replaced tobacco in the newsletter if not on farms.

The KDA continued to deploy the rhetoric of tobacco heritage throughout the 1980s. In January 1982, a full-page pictorial on tobacco ran under the title "'Bacca's in Case," including photographs

6--KENTUCKY AGRICULTURAL NEWS, JANUARY, 1982

'BACCA'S IN CASE

It takes a lot of work to get tobacco from the seedling stage to hanging in the barn, and even longer before finally getting it to market. In between, though, some people turn work into competition. Middle right, Jerry Bailey of Owensboro competes in a Woodford County championship tobacco cutting contest organized by Larry Roberts, middle left. Roberts is, by all accounts, the fastest tobacco cutter in the world. It's a fact that he's the only man known to beat a tobacco cutting machine.

"'Bacca's in Case." (*Kentucky Agricultural News*, January 1982, Courtesy of the Kentucky Department for Libraries and Archives)

of cutting tobacco, tobacco growing in the fields, and a tobacco warehouse. The use of *'bacca* and *in case* references both a common vernacular pronunciation of *tobacco* and the term used by farmers to refer to tobacco that is ready to be stripped; therefore the use of these terms provides yet another example of an appeal to farmers through the deployment of their own language and knowledge. The photo of tobacco cutting was taken at a cutting contest in Woodford County, marking the first mention of such a contest, followed by an August 1982 article about a tobacco festival and cutting contest in Georgetown. "The real purpose of the festival is to celebrate the importance of tobacco to Kentucky," according to its organizer.[39] Tobacco festivals had begun in different periods, serving different purposes. In the 1930s and 1940s festivals were put on in large tobacco marketing towns as a means of boosting warehouse business in hard economic times.[40] In the early 1980s, tobacco festivals were yet another example of the movement to celebrate tobacco as heritage. As I will discuss, such festivals later became a symbol of just how much things would change.

The tobacco "way of life" was celebrated through art yet again in October 1984, when Philip Morris unveiled an exhibit entitled *Twelve Months of Tobacco.*[41] Despite the fact that this series was described in the KDA newsletter as evidence of Philip Morris's significant role in the tobacco tradition, Philip Morris is not the subject of these paintings—farmers are. According to the article, the paintings provide an "accurate and moving portrayal of the month-by-month life of the Kentucky burley farmer."[42] This exhibit of twelve oil paintings by Kentucky artist Toss Chandler was first displayed in the Halls of Congress and later traveled around Kentucky. According to Kentucky senator Walter "Dee" Huddleston, the exhibit demonstrated "the significance of tobacco companies such as Philip Morris, to Kentucky and the entire nation. Philip Morris has been among the most civic-minded corporate citizens of Kentucky. The exhibit is just another step in that tradition. The exhibit also moves us to pay tribute to the thousands of hard working tobacco farmers—most of them small farmers in the purest sense of the word—who, down through the years, have provided so much to the country, the economy and the well-being of Kentucky."[43]

This series speaks to multiple audiences: to a general audience it argues that tobacco is about not only disease and addiction but also both hardworking farmers and American heritage. At the same time, like the Reynolds "Pride in Tobacco" campaign, *Twelve Months of Tobacco* deploys, yet again, what Burke calls "the simplest case of persuasion," "persuad[ing] a man only insofar as you can talk his language by speech, gesture, tonality, order, image, attitude, idea, *identifying* your ways with his."[44] By seeing themselves in such portraits, farmers were reminded that their interests remained aligned with those of the industry—and specifically with a company framed as "among the most civic-minded corporate citizens of Kentucky," a clear example of attempts by the industry to counter threats to its public image.

The industry also deployed the rhetoric of heritage with the farmer as a central character in other arenas at this time. According to a Tobacco Institute pamphlet, "Kentucky's Tobacco Heritage," "the cultivation, sale and manufacture of tobacco play vital roles in the history and economy of the Bluegrass State, as Kentuckians toil to nurture and harvest the fine, light leaf used around the world in cigarettes, smoking and chewing tobacco and snuff."[45] The centerfold of the pamphlet is a collage of photos of tobacco work from the period, including the quintessential photos of barns, fields, and tobacco being cut. This pamphlet is undated, but the photographs and the mention of baling tobacco near the end of the narrative suggest that it was published in the 1980s. It ends with an emphasis on tobacco farmers in the present. Under the heading "Carrying on the Tobacco Tradition" the final statement reads: "The hard work and enthusiasm of the thousands of Kentuckians involved with tobacco assure that the golden leaf will always be an important part of the life and economy of the Bluegrass State."[46] Also in the 1980s, R. J. Reynolds established the Tobacco Farm Life Museum in North Carolina, with the goal, as described by Peter Benson, of "espous[ing] a positive view of tobacco as heritage" as part of a campaign "to portray growers, and motivate growers to reckon themselves, as a group of people whose culture and tradition have been disrespected."[47] Tobacco art, published historical narratives, and museums are all examples of the deployment of the rhetoric of heritage as the tobacco industry came to understand that it was losing its public relations battle. Tobacco farmers were increasingly pushed forward as the face of the industry in attempts to induce empathy from tobacco's critics and to ensure that farmers themselves identified with the industry.

In April 1985, the "attack" rhetoric returned as growers and companies were together described as "fight[ing]" the Reagan administration's attempts to phase out the tobacco program.[48] The bulk of this front-page article turns out to be about the plans under way for a buyout of the pool stocks held by the co-op, which had again grown too large, this time in part due to the inferior-quality tobacco left over from the 1983 drought year. In his message, the commissioner described the proposed buyout in vague terms, contrasting it with a doomsday scenario in which, if the deal did not go through, "the destruction of the tobacco program would result in a degree of economic chaos in Kentucky unmatched since the Great Depression."[49] These two articles combined to create an argument that farmers must approve this buyout—which, not so incidentally, also included a reduction of the support price by thirty-three cents per pound—that relies on praising the efforts of farmers in the "fight" and instilling fear that if they did not continue the fight by voting for the buyout they risked losing everything. Farmers were successfully convinced that the 1985 buyout of pool stocks and reduction of the support price were the only way to save the program; the referendum passed.

Tobacco and alternative crops remained side by side in the newsletter's pages in this period. In July 1985, the KDA reported on farmers who were growing bell peppers with the encouragement and marketing assistance of the KDA. The following October, in an article headlined "1985 Was a Growers' Learning Experience," the KDA reported that its attempt to find "profitable alternatives to tobacco" had not proved as successful as farmers and the KDA had hoped.[50] The message was that farmers were looking at what they had done wrong in hopes of trying again next year, although clearly marketing—which the KDA had claimed as its expertise—was a key problem. Many growers were encouraged to raise bell peppers to supplement or replace their tobacco income in this period. As I will describe in a later chapter, this effort was nearly universally unsuccessful, and bell peppers are in fact the quintessential crop about which failed diversification narratives are told.

Throughout the 1980s, attacks on tobacco came in the form of attempts to dismantle the tobacco program, rather than in the form of tax increases, as was the case in the 1970s and 1990s. In October 1986, the commissioner of agriculture expressed his hope that 1987 would be a better year for tobacco farmers because "tobacco is what pays the bills . . . tobacco is something most farmers grow some of. Try as they might, opponents of tobacco will not in our lifetime succeed in doing away with the kind of program we need to allow this element of the Kentucky farm economy to survive."[51] In one of the most interesting tobacco displays over the course of the newsletter, this issue also included a full-page tobacco crossword puzzle with over two hundred clues, mostly tobacco related, leading to terms ranging from *tomahawk* (used to cut tobacco) to *blue mold* (a disease that can destroy a tobacco crop).[52] The puzzle might suggest that

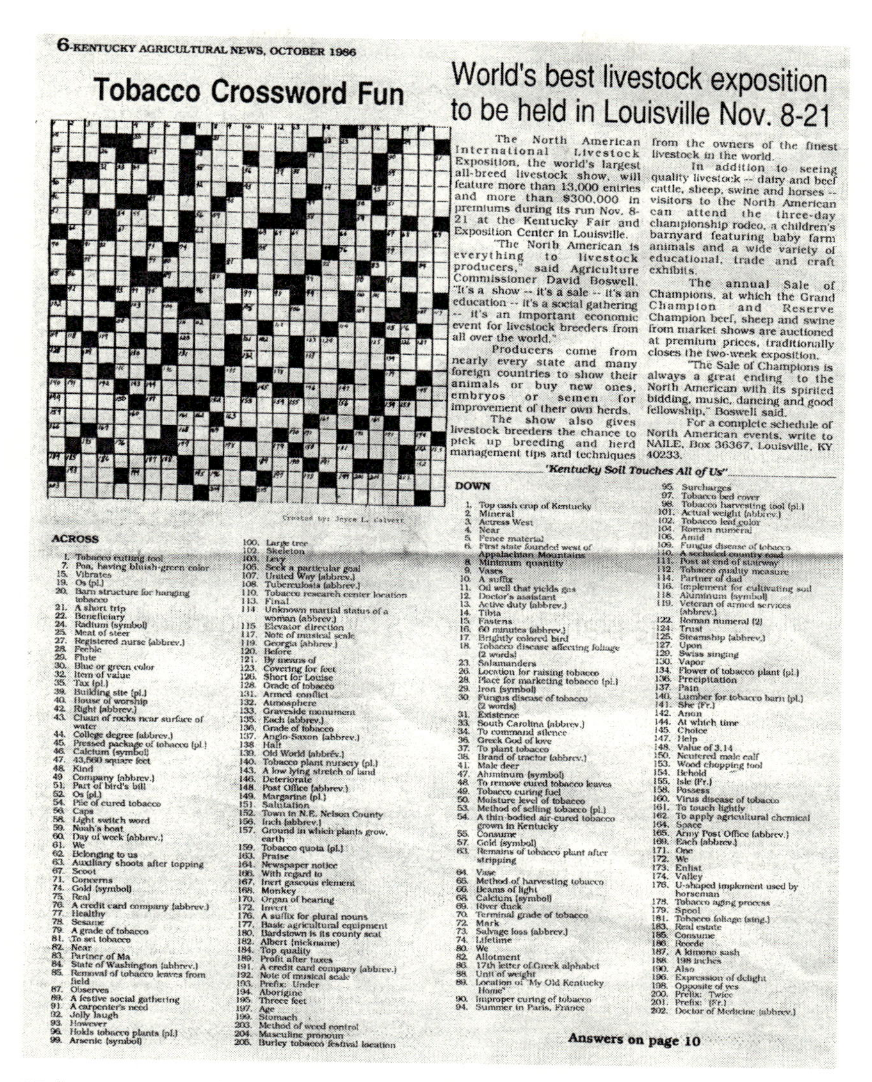

6-KENTUCKY AGRICULTURAL NEWS, OCTOBER 1986

Tobacco Crossword Fun

Created by: Joyce L. Calvert

ACROSS

1. Tobacco cutting tool
7. Poa, having bluish-green color
15. Vibrates
19. Os (pl.)
20. Barn structure for hanging tobacco
21. A short trip
22. Beneficiary
24. Radium (symbol)
25. Meat of steer
27. Registered nurse (abbrev.)
28. Feeble
29. Flute
30. Blue or green color
32. Item of value
35. Tax (pl.)
39. Building site (pl.)
40. House of worship
42. Right (abbrev.)
43. Chain of rocks near surface of water
44. College degree (abbrev.)
45. Pressed package of tobacco (pl.)
46. Calcium (symbol)
47. 43,560 square feet
48. Kind
49. Company (abbrev.)
51. Part of bird's bill
52. Os (pl.)
54. Pile of cured tobacco
56. Caps
58. Light switch word
59. Noah's boat
60. Day of week (abbrev.)
61. We
62. Belonging to us
63. Auxiliary shoots after topping
67. Scoot
71. Concerns
74. Gold (symbol)
75. Real
76. A credit card company (abbrev.)
77. Healthy
78. Sesame
79. A grade of tobacco
81. To set tobacco
82. Near
83. Partner of Ma
84. State of Washington (abbrev.)
85. Removal of tobacco leaves from field
87. Observes
89. A festive social gathering
91. A carpenter's need
92. Jolly laugh
93. However
96. Holds tobacco plants (pl.)
99. Arsenic (symbol)
100. Large tree
102. Skeleton
103. Levy
105. Seek a particular goal
107. United Way (abbrev.)
108. Tuberculosis (abbrev.)
110. Tobacco research center location
113. Final
114. Unknown martial status of a woman (abbrev.)
115. Elevator direction
117. Note of musical scale
119. Georgia (abbrev.)
120. Before
121. By means of
123. Covering for feet
126. Short for Louise
128. Grade of tobacco
131. Armed conflict
132. Atmosphere
133. Graveside monument
135. Each (abbrev.)
136. Grade of tobacco
137. Anglo-Saxon (abbrev.)
138. Halt
139. Old World (abbrev.)
140. Tobacco plant nursery (pl.)
143. A low lying stretch of land
146. Deteriorate
148. Post Office (abbrev.)
149. Margarine (pl.)
151. Salutation
152. Town in N.E. Nelson County
156. Inch (abbrev.)
157. Ground in which plants grow, earth
159. Tobacco quota (pl.)
163. Praise
164. Newspaper notice
166. With regard to
167. Inert gaseous element
168. Monkey
170. Organ of hearing
172. Invert
176. A suffix for plural nouns
177. Basic agricultural equipment
180. Bardstown is its county seat
182. Albert (nickname)
184. Top quality
189. Profit after taxes
191. A credit card company (abbrev.)
192. Note of musical scale
193. Prefix: Under
194. Aborigine
195. Three feet
197. Age
199. Stomach
203. Method of weed control
204. Masculine pronoun
206. Burley tobacco festival location

World's best livestock exposition to be held in Louisville Nov. 8-21

The North American International Livestock Exposition, the world's largest all-breed livestock show, will feature more than 13,000 entries and more than $300,000 in premiums during its run Nov. 8-21 at the Kentucky Fair and Exposition Center in Louisville.

"The North American is everything to livestock producers," said Agriculture Commissioner David Boswell. "It's a show -- it's a sale -- it's an education -- it's a social gathering -- it's an important economic event for livestock breeders from all over the world."

Producers come from nearly every state and many foreign countries to show their animals or buy new ones, embryos or semen for improvement of their own herds.

The show also gives livestock breeders the chance to pick up breeding and herd management tips and techniques from the owners of the finest livestock in the world.

In addition to seeing quality livestock -- dairy and beef cattle, sheep, swine and horses -- visitors to the North American can attend the three-day championship rodeo, a children's barnyard featuring baby farm animals and a wide variety of educational, trade and craft exhibits.

The annual Sale of Champions, at which the Grand Champion and Reserve Champion beef, sheep and swine from market shows are auctioned at premium prices, traditionally closes the two-week exposition.

"The Sale of Champions is always a great ending to the North American with its spirited bidding, music, dancing and good fellowship," Boswell said.

For a complete schedule of North American events, write to NAILE, Box 36367, Louisville, KY 40233.

"Kentucky Soil Touches All of Us"

DOWN

1. Top cash crop of Kentucky
2. Mineral
3. Actress West
4. Near
5. Fence material
6. First state founded west of Appalachian Mountains
8. Minimum quantity
9. Vases
10. A suffix
11. Oil well that yields gas
12. Doctor's assistant
13. Active duty (abbrev.)
14. Tibia
15. Fastens
16. 60 minutes (abbrev.)
17. Brightly colored bird
18. Tobacco disease affecting foliage (2 words)
23. Salamanders
26. Location for raising tobacco
28. Place for marketing tobacco (pl.)
29. Iron (symbol)
30. Fungus disease of tobacco (2 words)
31. Existence
33. South Carolina (abbrev.)
34. To command silence
36. Greek God of love
37. To plant tobacco
38. Brand of tractor (abbrev.)
41. Male deer
47. Aluminum (symbol)
48. To remove cured tobacco leaves
49. Tobacco curing fuel
50. Moisture level of tobacco
53. Method of selling tobacco (pl.)
54. A thin-bodied air-cured tobacco grown in Kentucky
55. Consume
57. Gold (symbol)
63. Remains of tobacco plant after stripping
64. Vase
65. Method of harvesting tobacco
66. Beams of light
68. Calcium (symbol)
69. River duck
70. Terminal grade of tobacco
72. Mark
73. Salvage loss (abbrev.)
74. Lifetime
80. We
82. Allotment
86. 17th letter of Greek alphabet
88. Unit of weight
89. Location of "My Old Kentucky Home"
90. Improper curing of tobacco
94. Summer in Paris, France
95. Surcharges
97. Tobacco bed cover
98. Tobacco harvesting tool (pl.)
101. Actual weight (abbrev.)
102. Tobacco leaf color
104. Roman numeral
106. Amid
109. Fungus disease of tobacco
110. A secluded country road
111. Post at end of stairway
112. Tobacco quality measure
114. Partner of dad
116. Implement for cultivating soil
118. Aluminum (symbol)
119. Veteran of armed services (abbrev.)
122. Roman numeral (2)
124. Trust
125. Steamship (abbrev.)
127. Upon
129. Swiss singing
130. Vapor
134. Flower of tobacco plant (pl.)
136. Precipitation
137. Pain
140. Lumber for tobacco barn (pl.)
141. She (Fr.)
142. Anon
144. At which time
145. Choice
147. Help
148. Value of 3.14
150. Neutered male calf
153. Wood chopping tool
154. Behold
155. Isle (Fr.)
158. Possess
160. Virus disease of tobacco
161. To touch lightly
162. To apply agricultural chemical
164. Space
165. Army Post Office (abbrev.)
169. Each (abbrev.)
171. One
172. We
173. Enlist
174. Valley
176. U-shaped implement used by horseman
178. Tobacco aging process
179. Spool
181. Tobacco foliage (sing.)
183. Real estate
185. Consume
186. Recede
187. A kimono sash
188. 198 inches
190. Also
196. Expression of delight
198. Opposite of yes
200. Prefix: Twice
201. Prefix: (Fr.)
202. Doctor of Medicine (abbrev.)

Answers on page 10

"Tobacco Crossword Fun." (*Kentucky Agricultural News*, October 1986, Courtesy of the Kentucky Department for Libraries and Archives)

knowledge of tobacco production was or should be common knowledge, or that tobacco knowledge was becoming specialized knowledge as the crop was increasingly threatened. Certainly, however, it is both an expression of self-consciousness of tobacco knowledge and a further claim to KDA's authority over this knowledge.

The KDA newsletter became a quarterly publication in 1983, and between that year and 1991 tobacco coverage of some kind was missing from only three issues. The openings of the markets and the tobacco exhibit at the state fair were inconsistently reported on, but heritage-based coverage of tobacco remained steady. Beginning in the late 1980s, tobacco was on the front page less and less frequently, and if it was there it was often below the fold. The need for farmers to diversify, particularly into food production, was the increasing focus, and the word *diversification* appeared more and more often.

In April 1989, the remarks made by a Philip Morris representative at a statewide agriculture conference were summarized. He had apparently given a mixed message to attendees, noting that the "next few years should be good ones for burley producers. But he warned that tobacco would continue to come under attack in Congress."[53] Quotas were in fact raised 24 percent in 1989, and it was feared that "underproduction" was making US burley growers seem undependable. In January 1990, Danny McKinney of the Burley Co-op was quoted as saying that although quotas had been raised by 24 percent, farmers had only planted 18 percent more.[54] He mentioned labor shortages as one cause of the shortfall, and indeed, in April 1990, the first mention of migrant labor as a solution to labor shortages appeared.[55] While the lack of labor would have been a major impediment to increased production, access to 24 percent more of all of the inputs farmers needed—from seed and fertilizer to equipment and land—would also have proven difficult for farmers to come by. This is a clear example of the way in which rising and falling quotas were used to keep farmers scrambling to meet the needs of the tobacco companies, and farmers were then consistently blamed for both over- and underproduction.

TOBACCO'S LAST STAND

In July 1990, the chairman of the legislative Tobacco Task Force, Representative Donnie Gelding, used the guest column of the KDA newsletter to describe "attacks" by the "anti-tobacco lobby," which was once again trying to raise taxes and destroy the tobacco program.[56] He noted that quota was up 2.5 percent in 1990 and stated, "This must come as quite a surprise to anti-tobacco zealots who are claiming that tobacco is a dying industry," when in fact the tobacco program, according to Representative Gelding, was the healthiest

Kentucky's Pride

The Official Publication of The Kentucky Department of Agriculture

October 1990 ISSN 0454-8604 Volume 21, No. 4

New program assures quality

Mexican ranchers look to Kentucky for livestock

By Millie Mattingly

As part of his program to develop international markets for Kentucky agriculture, Commissioner Ward "Butch" Burnette traveled to Mexico in September for a week of meetings with Mexican officials, cattlemen and ranchers. Director of International Marketing Hans Petereit and Deputy Commissioner Roy Massey accompanied Burnette.

Discussions at these meetings centered on increasing agricultural trade between Kentucky and Mexico. It is expected that a memorandum of agreement between the two will be

"Kentucky is a major livestock-producing state. While we have exported large numbers of livestock to Mexico in the past, we want to develop programs that will give Kentucky a marketing advantage over other states and help our producers capture a bigger share of the livestock export market," he said.

While in Mexico, Burnette visited several ranches and listened to a group of over 50 cattlemen and ranchers talk about their needs and problems they've had importing livestock from the U.S.

"The department is currently working on a program for marketing

Photo by Millie Mattingly

Fall Scene - Kentucky Style

Tobacco harvesting has moved into curing and stripping season. Tobacco markets open November 19. Pounds this year are expected to be 17 percent above last year's crop and the quality appears to be good.

"Fall Scene—Kentucky Style." (*Kentucky's Pride,* October 1990, Courtesy of the Kentucky Department for Libraries and Archives)

of all federal agriculture programs.[57] This remark now reads like foreshadowing of what was to come in the next two decades, as the tobacco program ended and tobacco came to be seen as dying if not dead. By 1990 the days of editorializing commissioners were largely over, even the rhetoric of tobacco as heritage had become increasingly rare, and the voices of Gelding and a handful of others represented the dwindling of outspoken tobacco advocates in the pages of the KDA newsletter.

By this time, there was rarely a direct reference to health issues; that battle seems to have been accepted as lost. The number of articles that covered self-evident aspects of tobacco production sharply declined. An October 1990 cover photo of a scene from a tobacco harvest, with the caption "Fall Scene, Kentucky Style," represents one of the last images that unquestioningly presents that *what fall means in Kentucky is tobacco.* Elsewhere in the issue is a photo of Bobby Preston, winner of the Garrard County Tobacco Cutting Contest for the ninth year in a row.[58]

Senator Ford wrote a guest column in January 1992, entitled "Tobacco Program Is Still Vital to State," in a tone bordering on resignation. Ford wrote, "I took pride in my [tobacco] crop. . . . I know there are thousands of Kentucky farmers experiencing that same justified pride as they take their burley crop to the market," and "we have fought hard for our burley program and I believe that every person involved in the industry can take pride in the strides that have been made. Now is not the time to let our guard down."[59] Here the fight was starting to sound like it had taken place in the past—he *took pride* in his crop, and we have *fought* hard—as though Ford were trying to keep the soldiers in the fight for tobacco from fleeing the battlefield: "Now is not the time to let our guard down."

However, the fight continued in other publications, such as two books published in the 1990s by major tobacco organizations, the Burley Auction Warehouse Association and the Burley Tobacco Growers Co-operative Association. Both of these provided histories published in commemoration of the organizations' respective fiftieth anniversaries, but they each also provided the now-familiar historical narrative. According to the Burley Tobacco Growers Co-operative Association, "Jamestown, in the Virginia Tidewater, was a miserable place on a swampy island that took the lives of 800 among the first 1,000 settlers in its first four years of existence. It was against this background of starvation and misery, then, that the American tobacco industry began."[60] According to the Burley Auction Warehouse Association, "Selling tobacco leaf . . . is the basis for a time-honored culture. The history of this commerce interweaves with America's development, as solidly as do the culture-thick commerces of the industrial revolution, the cattlemen of the Old West, or the modern automotive industry."[61] Each organization used tobacco heritage as a means to promote its own relevance on the anniversary of its establishment.

President Bill Clinton provided the fodder for one of the last "fights" for tobacco that the KDA gave attention to in its newsletter, although this one was fought in milder terms. In April 1993, the KDA noted that the previous burley market had been "one of the best in history in terms of price and volume" (and instead of a reward growers would face a 10 percent quota cut in the coming crop year), but there was an upcoming fight against tobacco tax increases that, according to the commissioner, was going to be difficult to win.[62] The commissioner sounded resigned to tax increases, as he wrote about his hope that policy changes would be put into place, linked to the tax increases, to help tobacco producers.

In January 1994, Commissioner Ed Logsdon devoted his column to the power that tobacco companies maintained over farmers:

"Ever wonder why tobacco farmers seem to be in a continual state of confusion and frustration? Maybe it's because the tobacco companies want it that way."[63] According to the commissioner, tobacco farmers had been told they were not producing enough burley just a few years ago, so they stepped it up, and now they were being told they were overproducing. They had been told to stop stripping into just one grade, and they started stripping into three; now they were getting paid just about the same price for all grades. The previous year the companies hadn't complained about the price, and now they were claiming that "U.S. tobacco is terribly overpriced in the world market and that farmers must accept a lower price in order to compete."[64] He went on to argue that the companies had convinced the farmers to protest the tax increase, and that while he agreed that they all needed to unite and fight against the Clinton tax increase, he wanted to see the companies make a commitment to growers.

This explicit critique of the tobacco companies is telling of how times had changed, as Logsdon's words signal a rhetorical shift in the alignment of the KDA away from the tobacco companies. By the mid-1990s, public opinion about tobacco manufacturers had changed markedly, a fact that would have made it extremely advantageous for the KDA to distance itself—and farmers—from the companies. In fact, several occurrences in 1994 radically changed the tide of public perception of the tobacco industry. These included an ABC News feature that exposed the practices of the industry in controlling the amount of nicotine in cigarettes (increasing the levels of nicotine in order to replace nicotine lost from lowering the amount of tar) and the televised testimony of seven industry leaders before a congressional subcommittee.[65] Before an audience of CNN and C-SPAN viewers, the seven CEOs all proclaimed that they believed that nicotine was not addictive. It was at this time that "an altogether different narrative of smoking began to emerge: one of secret technologies, research laboratories precipitously closed down, and nicotine as a potent drug carefully added to a high-tech product to sustain a greedy industry's profits."[66] In this context, it is hardly surprising that the KDA would distance itself from the industry with which it had previously aligned itself. This need for repositioning, combined with the growing resignation that tobacco was going to be less important in the future because it could not escape stigma, declining markets, and other challenges, was behind Logsdon's critique—however accurate. Although the KDA had not yet abandoned tobacco altogether, its advocacy became increasingly tempered by an acknowledgment of anti-tobacco arguments.

In April 1994, the commissioner returned to the defense of tobacco against the attacks of the Clinton administration, arguing that it was trying to tax tobacco out of existence. He also introduced a preamble to his argument that articulated a new position on tobacco use for the KDA: "I do not advocate smoking, but I am a strong believer that public policy should be based on facts, not emotion. Anti-tobacco forces are seeking to eradicate tobacco by making cigarettes unaffordable and unavailable."[67] People have the right to make informed decisions about smoking, he argued. He noted in this same column—without observing the irony—that recent state legislation had required the KDA to enforce underage smoking bans. He informed readers that "we did not seek this responsibility, but I want all Kentuckians to know that we will take this seriously."[68]

In July 1994, Governor Brereton Jones guest authored a piece entitled "Fairness Essential to Tobacco Growers," in which he argued against the proposed federal tobacco tax increase: "As a tobacco farmer myself, I realize how important tobacco is to our state's economy. My wife and I raise over 150,000 pounds every year, and we depend on the income we generate from this crop."[69] He argued that if the tax was going to be increased, tobacco farmers should be compensated for the lost income that would result. This period marks the beginning of discussions about farmer compensation for lost income, an idea that became a reality later with the Master Settlement Agreement and the tobacco buyout. It can also be read as a sign of resignation: the governor would continue to fight for tobacco farmers, but he was realistic enough to see that the end was coming: "We must continue to fight to see that tobacco companies use homegrown, rather than imported leaf"; "If we do not, we will wake up one day, our allotment will be gone, and we won't have anything to show for it."[70] Jones was the last Kentucky governor who was also a tobacco grower.

In January 1996, incoming commissioner Billy Ray Smith (a tobacco farmer) also sent a message that combined a vow to fight for tobacco with a subtext that implied that tobacco was fading, stating that he "will not forget tobacco. 'I will support tobacco wholeheartedly,' he said. 'If people want tobacco, we should be growing it and getting money for it. It's part of our heritage.'"[71] By vowing not to "forget" tobacco, he implied that it was fading from view—becoming heritage. In May 1996, the KDA proclaimed that Commissioner Smith had taken the lead against teen smoking—a very different attitude from his predecessor's grudging acceptance of this new responsibility and a very long way from the commissioners of decades earlier, who had applauded increases in smoking rates as good for Kentucky's economy. By this time lawsuit after lawsuit had been filed against the industry, but each had failed. In 1997 the industry would settle such a case for the first time.[72]

An "Uncertain" Future

The kinds of articles that had once been obligatory began to fade in the late 1990s. The winter 1998 issue of the KDA newsletter included the classic photo of the commissioner at a tobacco warehouse on the opening day of the marketing season. In the photo, however, Commissioner Billy Ray Smith suggests tobacco heritage by wearing a trench coat and a fedora hat, the garb of his 1940s and '50s predecessors. There was no opening day photo in 1999, but it returned in 2000. In marked contrast to the attention paid to matters of tobacco policy in earlier decades, the 1998 Master Settlement Agreement between the state attorneys general and the four major tobacco companies was not mentioned until January 1999. Coverage of this historic lawsuit was limited to discussions about how Kentucky's share of the funds might best be used to the benefit of farmers and for the promotion of the agricultural economy.

By the late 1990s the rhetoric had shifted to the "uncertain" future of tobacco. The uncertainty became explicit in the fall 1998 issue, eventually replacing the rhetoric of attack until tobacco had largely faded from the pages of the KDA newsletter. This shift from "attack" to "uncertainty" appeared, for instance, in an article on the importance of agriculture to Kentucky: "However, recent concerns over the political fate of the tobacco industry and the economic weakness of crop and livestock industries have left Kentucky farm families uncertain about the future."[73] The rhetoric of uncertainty picked up steam in April 2000 with a front-page article: "What's Next for Tobacco? Many Issues Unresolved after Devastating Winter." According to the article, "As this newspaper went to press, Kentucky tobacco farmers were facing more uncertainty than they had encountered since the price-support system was adopted some 60 years ago. Several issues remained unresolved on the eve of tobacco setting time."[74] This uncertainty revolved around how MSA funds would be used, the beginning of direct contracting between farmers and tobacco producers, and quota cuts at the astounding rate of 45.3 percent: "Critics of the plan [by Philip Morris to contract directly with farmers] charged that it was the first step toward eliminating the price support program."[75] Indeed it was.

The July 2000 issue featured a lengthy article about three tobacco farmers and the decisions they were making about the future, in a front-page piece entitled "2001: A Farm Odyssey." One farmer, Roger Quarles (who later became president of the Burley Co-op), intended to keep raising tobacco and had just built a new tobacco barn before the massive quota cut. Quarles provided a perspective on the search for alternative crops that continues to be expressed by many growers today: "I couldn't see not growing tobacco in favor of growing something in the alternative category that may or may not work. I don't see the smartness in quitting a high-income crop like tobacco as long as it is available. Our problem in trying to think of something else is that you have to have a consistent income. Bills are consistent. The mortgage is consistent."[76] He then pointed out the irony that "we always used tobacco income as a subsidy for trying something new," making it even more risky to try new things in this time of shrinking quotas. Quarles went on to say that he thought the tobacco companies were purposefully not buying American tobacco: "The tobacco companies' buying decisions today 'are not based on traditional ways of doing things, or else they'd be purchasing more.'" He concluded, "I don't think we'll ever lose tobacco. I think we'll always grow tobacco. But there's going to be some major changes."[77]

The two other farmers featured were working to build vegetable co-ops but also continued to raise tobacco, suggesting agreement with Quarles's argument that tobacco income remained a safety net. One of them, Paula Franke, described "Philip Morris' plan to contract with some growers, [as] 'akin to union-busting,'" saying about the company, "They're not killing a way of making a living. They're killing a way of life." She was "pessimistic about the future of tobacco in Kentucky. 'I think it's going to be the end,' she said. 'I think this is the last year we'll see tobacco production as we have known it. I think it's over.'"[78] Here she described tobacco as a way of life that was ending.

From this point forward, tobacco was rarely the exclusive subject of articles and was mentioned almost entirely in passing. In January

Kentucky AGRICULTURAL News

July 2000 — Published by the Kentucky Department of Agriculture — ISSN 1062-5836 — Vol. 32 No. 3

INSIDE

Agriculture Commissioner Billy Ray Smith

New ag board will help bring change to Kentucky.

Page 4

Research center marks milestone

The 75th anniversary of the University of Kentucky's Princeton research and education center will be celebrated at the UK College of Agriculture's Field Day July 20.

Page 2

Growers upbeat about farmers' markets

Fresh Kentucky produce is available at farmers' markets all over the state. In Lexington, producers also offer meat, eggs and dairy products.

Page 3

2001: A Farm Odyssey

This past spring, Kentucky Agricultural News *spoke to three tobacco farmers about their operations and the future of Kentucky's No. 1 cash crop. Their farms vary in size and are located in different sections of the state. These are their perspectives.*

The year 2001 was immortalized in a science-fiction movie a generation ago. To many, it is the true beginning of the new millennium.

To Roger Quarles, 2001 will be decision time for him and many of his neighboring Bluegrass farmers after they've sold this year's much smaller tobacco crop. Some will have to decide whether to raise another crop or get a job to supplement their tobacco income. Some may have to decide whether to get completely out of farming.

"I think it's going to be this coming year before anyone has to make a decision about what they're going to do," Quarles said in late May on his Scott County farm.

Quarles is farming his own 300 acres but, for the first time in his career, isn't sharecropping this year; the Franklin County farm that he used to work was sold for development, he said. Due to a quota carryover, Quarles is actually growing more tobacco this year – about 34 acres, or

See QUARLES, page 8

Ted Sloan

Roger Quarles built the tobacco barn at left just before burley tobacco quotas were reduced by 45 percent. He won't feel the full impact of the cut until 2001.

Burley cuts spur farmers to open vegetable co-op

Daviess County farmers have talked for years about starting a vegetable produce co-operative. So when burley tobacco quotas were cut by 45.3 percent last February, that was all the incentive many of them needed to take the co-op idea past the idea stage.

One of those farmers is Joe Elliott, a member of the board of the West Kentucky Growers Cooperative. Elliott saw his tobacco acreage cut nearly in half, from 110 acres last year to 61 this year. His greenhouse business, which grossed $100,000 last year, managed only $40,000 this year.

See ELLIOTT, page 8

Lewis County growers start a farmers' market. Page 9

"2001: A Farm Odyssey." (*Kentucky Agricultural News,* July 2000, Courtesy of the Kentucky Department for Libraries and Archives)

2001, the photograph of the opening of the markets once again appeared, with the headline "The Tradition Continues."[79] The obvious reference is to ongoing auctions despite the first year of Philip Morris's direct contracting program, but the subtext is debates about whether contracting would bring the end of the warehouse system. In this same issue, the commissioner mentioned—in the manner of an afterthought at the end of his message—the upcoming referendum on the continuation of the program: "Also, tobacco farmers should be sure to cast their ballots in the February election on whether to continue the federal tobacco quota and price support program."[80] In marked contrast to past commissioners' messages, he offered no opinion on how growers should vote.

The "uncertainty" rhetoric continued in October 2001, with

tobacco mentioned in passing in an article about KDA's efforts to "boost" vegetable cooperatives: "With the changing face of tobacco and lower commodity prices for traditional crops, Kentucky's farmers are actively seeking ways to add value to their products and diversify their operations."[81] The January 2003 issue included a front-page auction photo with the title "Auctions Survive—for How Long?" and a brief caption describing the burley market, in contrast to market updates in the past that took up one or more lengthy articles.

Tobacco heritage returned in July 2003, in a front-page article about the winning photographs in a KDA photography competition: "Photos Reflect Past, Future of Ky. Agriculture."[82] What counted as *past* and what counted as *future* were explicitly noted. The winning photo depicts a window through which tobacco can be seen hanging in an old structure. Weeds grow tall in front of it, and a wooden beam crosses diagonally through the center of the photo—a barrier between the viewer and this glimpse into the past. The photo was chosen "for its artistic quality and the quiet symbolism of the *history* of tobacco in the Commonwealth."[83] In contrast, the second-place photo features a three-year-old boy waving at his grandfather as he cuts hay: "Judges said the photo showed hope for the *future* of the family farm."[84] Tobacco had been completely recategorized as a symbol of Kentucky's past, an argument that was made visually as well as textually. In contrast to past tobacco "art," this photo honoring tobacco does not feature the action of tobacco work but a passive image of tobacco hanging in a barn nearly subsumed by weeds.

In October 2003, Commissioner Smith wrote his goodbye, mentioning tobacco once: "Much has been done but the future leaders of Kentucky agriculture still have much to do. As tobacco's future remains uncertain, we must sustain the momentum we have established."[85] This was followed by Richie Farmer's first commissioner's message in January 2004, which included a brief mention of tobacco: "There can be no doubt that Kentucky's agriculture community faces many challenges. Tobacco farmers are moving to diversify their farm operations."[86] This statement implies that all tobacco farmers were diversifying away from the crop, and it reflects the message that was to become central to Farmer's administration and to the state's overall message: Kentucky was becoming the home of a diversified agriculture economy. This issue included a photo of a "Silent Auction," a heading with a dual meaning: the direct meaning is the handheld bidding machines being used by buyers, replacing the classic chant

Kentucky
AGRICULTURAL
News

July 2003 | Published by the Kentucky Department of Agriculture | ISSN 1062-5836 | Vol. 35 No. 3

INSIDE

Agriculture Commissioner Billy Ray Smith

Ag Department takes center stage at the state fair.
Page 4

Marketing campaign gears up
Page 3

KDA directory ... 4
Short Rows ... 10
Classifieds ... 11

PERIODICAL POSTAGE PAID

Archival Copy

KyP
354.5
A2787ag

The winner!

Thomas G. Barnes

This photo was judged the best of more than 2,000 photos entered in the Kentucky Department of Agriculture's photo contest. Story and more photos, pages 7 and 12.

Cookbook promotes agri-tourism

By TED SLOAN
Kentucky Agricultural News

Kentucky's great foods, rich rural heritage and fun destinations will be showcased in a book due to be unveiled at the 2003 Kentucky State Fair.

"Pride of Kentucky: Great Recipes With Food, Farm and Family Traditions" is a coffee-table size cookbook filled with recipes using traditional Kentucky commodities. It contains stories about Kentucky's food, culture and traditions, and information on festivals and agri-tourism destinations in the Commonwealth. It is published by the Kentucky Extension Association of Family and Consumer Sciences.

"This book gives readers a taste of great Kentucky Fresh foods while also bringing them closer to the people who produce our food and fiber," Agriculture Commissioner Billy Ray Smith said. "The Kentucky Department of Agriculture is pleased to help the University of Kentucky Cooperative Extension Service market this book."

"'Pride of Kentucky' is a cookbook, but it is much more than that," said Scott Smith, dean of the University of Kentucky College of Agriculture. "It also contains fascinating lessons about Kentucky history, geography, sociology, and agricultural production. We are working on an educational curriculum that will utilize the book. It will be a tremendous resource for cooks, teachers, and people just looking

See COOKBOOK, page 6

"The Winner!" (*Kentucky Agricultural News*, July 2003, Courtesy of the Kentucky Department for Libraries and Archives)

of the tobacco auctioneer; the heading, however, also suggests the dying auction system, a system that would soon fall silent with the end of the program. There was no photo of the last auction season, for the sale of the 2004 crop.

In July 2004, tobacco's presence at the state fair returned to the newsletter, but not the tobacco exhibits—they were not mentioned, although they still take place. Instead, the passing reference provides more evidence of tobacco's recategorization into a crop of the past: "The KDA will join in on the historical theme of this year's fair. In the South Wing, the Department will celebrate the history of Kentucky agriculture with a mock tobacco barn, antique farm equipment and other displays alongside the fair's historical education and culture booths."[87] This equation of tobacco barns and antique farm equipment suggests that the barns have also lost their function.

The 2004 tobacco buyout, widely touted as "one of the most dramatic changes in any U.S. agricultural policy over the last half century,"[88] went almost entirely unmentioned in the pages of the KDA newsletter. There was never an article about the buyout, and the commissioner only mentioned it in passing in his January 2005 message: "We face some serious challenges in the months and years to come. With the tobacco quota buyout, the state's rural economy will never be the same"[89]—a statement that only served to emphasize the odd fact that the newsletter did not report on the buyout. Mention of the buyout after this was almost exclusively made in reference to the effect of the buyout payments on total farm receipts.

Although tobacco was rarely featured in the pages of the KDA newsletter by 2008, it continued to appear in lists of crops grown by farmers; lists of topics at Extension Service field days; farm receipt statistics; and, most often but indirectly, discussions of the tobacco settlement and buyout funds. Tobacco as a self-evident portion of the whole picture of Kentucky agriculture may be returning, to a degree, now that the wars on tobacco are understood to be lost, and heritage rhetorics have been largely abandoned by the KDA. At the same time, ongoing discussion of the uses of the settlement funds serve as a reminder of tobacco's importance in the past, as such articles focus on projects that are being carried out with the use of the funds, and these funds cannot be used for tobacco production—only for "diversification." I will return to this topic in a later chapter.

Tobacco Heritage Today

Although the KDA has largely abandoned the rhetoric of tobacco-as-heritage, this rhetoric has become fully embedded in the public discourses surrounding tobacco, from public art to media coverage. For instance, a mural on one section of the flood wall in Maysville—a northern Kentucky town on the Ohio River that was once a major

Tobacco is presented as part of the distant past on the flood walls in Maysville, Kentucky. (Photo by author)

tobacco marketing center—shows a frequently reproduced image: a farmer leaning on the wide door frame of a tobacco barn, looking out across his tobacco fields and additional tobacco barns. The viewer, with the farmer, looks out over the tobacco work happening in the distance. Although this image portrays work that may be happening just outside of town today, the context in which this mural exists tells a story of tobacco heritage as history. This is one of ten murals painted in a project in which "history becomes art," begun in 1998 to "visually demonstrate the historical connection of the Ohio River to Maysville."[90] The other panels include depictions of Indians hunting buffalo in deep snow, a nineteenth-century frontier settlement, a steamboat being loaded with hogsheads of tobacco and other goods, "the Marquis de Lafayette" being "Received at Maysville" in 1825, and slaves escaping along the Underground Railroad. Putting aside the tobacco mural, the image in this series that comes closest to a contemporary portrayal celebrates Rosemary Clooney, born in Maysville. Rosemary Clooney died in 2002, so even this most contemporary image mourns the passing of a local icon. The tobacco mural, understood within the context of the other murals that surround it, suggests that tobacco is as distant in time as buffalo roaming the rolling (and snowy) hills of Kentucky and as dead and gone as the celebrated songstress.

Tobacco festivals provide another telling example of the passing of tobacco, as they have now largely disappeared or their names have been changed. In 2001, the Logan County Tobacco Festival, which first took place in 1941 and 1942 and was then revived in 1957, became the Logan County Tobacco and Heritage Festival. This festival is unique in that it has retained direct associations with tobacco production, including a judging contest in which local farmers display their cured tobacco. The organizers of other festivals have made very different decisions. The Carrollton Two Rivers Tobacco/Fall Festival, which began in 1934, has become the Two Rivers Fall Festival, and the Washington County Sorghum and Tobacco Festival has been replaced by the Kentucky Crossroads Harvest Festival.[91] In 2006 and 2007, the Garrard County Tobacco Festival transitioned from the Tobacco Festival to the Tobacco/Rural Heritage Festival and finally to the Rural Heritage Festival. In these examples, either the word *tobacco* has disappeared altogether, or the word *heritage* has joined or replaced it.[92]

Former Garrard County agricultural extension agent Mike Carter, in an interview with me, described some of the history of the Garrard County Tobacco Festival. One attempt to change the name was made in the 1980s, but it was opposed by the community and failed. The renaming of the festival as the Rural Heritage Festival was approved by the festival committee amid grumbling from some community members but in the absence of outright public opposition. Carter views this as a symbol of "how things have evolved in tobacco country." The Garrard County Tobacco Cutting Contest, a wholly separate annual event drawing several hundred people, continues each year—the last remaining contest of its kind in Kentucky—despite growing opposition. Carter remarked, "Of course I've been asked, 'Well is this gonna be the "Ag Heritage Cutting Contest"?'"

It is important to note that Carter and others who object to such a name change are most likely responding to the removal of *tobacco* from the name rather than to the insertion of the word *heritage*. Yet these two processes are inseparable, a point that is made apparent through the comment about the cutting contest: Carter was not asked if *tobacco* might be dropped from the name; he was asked if *heritage* would replace it. It is doubtful that Carter and others who oppose the renaming of the festival would disagree with the idea that tobacco is part of Garrard County's heritage. The objection is instead to tobacco's erasure from Garrard County's present-day life and economy, its relegation to the past because tobacco is a crop that many in the community would prefer to ignore.

News coverage related to tobacco farming is laden with the rhetoric of death and dying. According to a May 2005 article in the *Lexington Herald-Leader*, "It's the dawning of a new and confusing world for Kentucky farmers"; an agricultural extension agent is quoted as saying, "You could be talking about the death of a culture."[93] A December 2005 headline engages this rhetoric even more literally, declaring, "One Mourner, but No Prayer for Tobacco, Burley Sale Becomes a 'Funeral' One Year after Quota Buyout."[94] This article uses the metaphor of a funeral, throughout, to describe the experiences of one woman, Pat Thompson, who is meant to represent all Kentucky tobacco growers, as she sells her tobacco on contract rather than at auction for the first time.[95]

This article presents not only the death of the tobacco auction but, synecdochically, just as Pat Thompson represents all farmers, the auction symbolizes the entire way of life of the tobacco family. Thompson married a tobacco farmer and lived her life by the "thirteen-month" cycle of the tobacco crop. Pat Thompson's way of life is presented as one that recalls the ways of the past: "She learned to cook on a wood stove in Shelby County. . . . She remembers when the electricity first came on. She remembers her house had the county's first phone because her daddy was a game warden. She remembers their first TV and how everybody took turns watching a Kentucky basketball game on the little 7-inch screen." The way of life of all the Pat Thompsons across Kentucky is here presented as one of life before electricity, phones, television, and other such symbols of modernity, but according to Pat Thompson, it was now "the end of an era." This article exemplifies the rhetoric that surrounded the situation for tobacco farmers in Kentucky and other tobacco-producing states by 2005. What was being mourned was not only the end of tobacco but what was understood to be the end of a way of life. Tobacco was now as much a part of the past as life without the modern conveniences of electricity and television.

While growers and many others continue to identify tobacco farming as an important source of income, public discourses have rhetorically recategorized the crop, the occupation, and the way of life as "heritage," devoid of economic importance in the present. The festival renamings and articles relegating the crop to the distant past are examples of a much larger shift in the public recognition of tobacco's role in the agriculture economy. As *heritage* either becomes the terminology used to describe tobacco farming or literally replaces it in public discourses, the term serves as a screen through which the public views tobacco farming. When viewed through the terministic screen of *heritage*, tobacco is in the past, and tobacco farmers are living relics of Kentucky's history.[96] Worse, they are growing a crop that is no longer understood to be profitable, and therefore those who still farm it are understood as simply too stubborn to let go of a dying tradition.

Tobacco heritage underwent an evolution that coincided with external events and changing attitudes. In the 1970s and 1980s, in the aftermath of the 1964 surgeon general's report and its consequences, both tobacco companies and the state celebrated tobacco heritage as an occupational way of life that was threatened. By the 1990s, it was becoming clear that heritage was not capable of recovering tobacco; it could neither salvage the public's opinion nor serve as a defense in the many lawsuits brought against manufacturers. During the 1990s, tobacco organizations published historical accounts of their former roles, tobacco art and displays focused on a distant past, the very word *tobacco* was replaced or joined by *heritage* in local festival names, and issues of current relevance to the tobacco industry disappeared from the pages of the Kentucky Department of Agriculture newsletter. Tobacco was erased from the present as "heritage," which now meant history.

According to James F. Abrams, "If textualized and thematized space is frozen into images of the past, people within the frame become actors, objects of memory, spectators to their own history."[97] Herein lies the danger of the heritage label, as applied to cultural practices that continue to be of economic importance to people in the present: individuals are objectified and their practices understood as outmoded. Not only do institutional expressions of tobacco as heritage erase tobacco farming and the tobacco farmer from the present, but they make tobacco farming *about* heritage and *not about* economic survival. However, the use of heritage discourses in noninstitutional contexts—by those in agricultural communities—can be read to offer a slightly different argument. They instead maintain that tobacco farming is an inherited way of life that continues to be of economic importance, one that has allowed Kentucky to hold onto small family farms while other regions have lost them. Farmers and extension agents frequently bring up the historical importance of tobacco, often listing details familiar from published historical narratives of tobacco, such as that tobacco was once used as currency, that it funded the Revolutionary War, and that tobacco leaves adorn our nation's Capitol. Garrard County tobacco producer G. B. Shell instructed me on tobacco history in a March 2007 interview:

> Author: What other things do uh, do people who don't know about tobacco—what other things do they need to know?
> Shell: Well tobacco has been important since the founding of this country. I guess a lot of people don't know that but around the Capitol of the United States, what do you see? Tobacco leaves, around the uh, the uh Capitol building. Do you know why they're there?
> Author: No sir why are they there?
> Shell: You don't know why they're there? In the . . . Revolutionary War, we borrowed money from France, to finance the war. And we gave them a mortgage on the tobacco crop. That's why they're there.

Mr. Shell and I had almost the exact same exchange in a January 2008 interview, except this time he teased me because he knew that I knew the answer. I told him I wanted to hear him tell it, and he did, but this time he added the coda, "Whenever these people put down tobacco so bad, they're putting down the whole country." G. B. Shell and others argue that although tobacco may be stigmatized in the present, we cannot forget what it did for *all of us* in the past; patriotism demands it.

The newsletter of the Kentucky Department of Agriculture provides a view of the changing political context of tobacco over the second half of the twentieth century and into the twenty-first, demonstrating that despite its changing status, tobacco has long been a contested crop with significant economic and symbolic value. Even though cattle (both dairy and beef), hogs, forages, and other farm products combined are featured in the newsletter far more frequently and consistently than tobacco, the coverage of tobacco demonstrates a self-consciousness that is absent from the coverage of all other farm products. It was tobacco that was featured on inaugural floats (along with thoroughbreds and bluegrass) and in commissioned art, not cattle, corn, forages, or other agricultural products that have risen and fallen in importance over the years. Such farm products have remained a self-evident part of the agricultural economy, rarely celebrated, while tobacco has long been central to a self-conscious presentation of Kentucky agriculture and identity, even with shifting meanings.

As the official voice of Kentucky agriculture, the KDA provides readers with "news" of the present state of agriculture; farmers and nonfarmers alike are asked to see this news as the current state of Kentucky agriculture and to imagine themselves as participants in the agricultural landscape. In the 1940s and into the 1950s, this meant an acceptance of tobacco as a vital agricultural commodity and of growing rates of smoking as good for Kentucky. With increased attention to the health consequences of tobacco use in the 1950s and 1960s, the Kentucky Department of Agriculture became a vocal advocate for the industry, deploying a rhetoric of tobacco's economic importance in defense of the crop and the products produced with it. Much of the rhetoric of this period was certainly political, meant to influence the future actions of readers, yet much of it can be understood as attempting to persuade readers that tobacco should be celebrated as an economic necessity in the present and, through the Burley Belle and similar pageants, as symbolic of Kentucky culture—not just Kentucky *agri*culture.

By the 1970s, the celebratory tone had receded somewhat as the KDA and its guest columnists became increasingly and more aggressively defensive in light of the "attacks" on tobacco from multiple fronts. Readers were to understand these attacks on tobacco as attacks on Kentucky more generally; as the commissioner of agriculture wrote in November 1976: "We all live better in Kentucky because of tobacco."[98] Yet it was in this period that the economic defense was joined, and eventually replaced, by heritage as a defense of tobacco, with an increasing focus on tobacco's importance in Kentucky's history that lasted into the 1980s as health consequences became impossible to deny. Rather than continue to deny tobacco-related disease, the KDA, in alignment with the tobacco companies, refocused its audience on tobacco as heritage with attempts to persuade readers that because tobacco had such historical importance, it should continue to be valued, and the life and work of the tobacco farmer should be celebrated as a way of life.

At the same time that the KDA presented a shifting self-consciousness about tobacco, in the pages of the newsletter certain self-evident features remained fairly constant until the turn of the twenty-first century. New technologies and farming practices were the most consistently covered topics throughout the newsletter's existence, along with articles about the tobacco program and other matters of policy. The commissioner's message continued as a regular feature of the newsletter, but under recent commissioners it has served almost exclusively as a celebratory site in which the work of the KDA is highlighted. A particular focus is agricultural diversification and the promotion of KDA's "Kentucky Proud" program, which encourages consumers to buy from Kentucky farmers and food producers. The KDA now presents a picture of the Kentucky agricultural economy as centered on alternative farm products such as vegetables, aquaculture, wine, and agritourism—despite farm receipts and census data that suggest a different picture, as I describe in a chapter to follow. The KDA asks farmer readers to imagine themselves in the "exemplary stories of praiseworthy people [who have successfully replaced tobacco with other farming opportunities]—and in their image those addressed are invited to make themselves over."[99]

The significant decrease in self-evident tobacco coverage over the last decade is as important as a consideration of the type of news that the KDA covers and how it covers this news. By the 1990s, the KDA seemed to find it politically expedient to disentangle itself from tobacco. Tobacco-as-heritage had largely disappeared from the newsletters by the late 1990s, and the details of major events such as the Master Settlement Agreement and the tobacco buyout—which meant the end of the Depression-era program and drastic changes to the farm economy—were not reported on. With the sharply diminished coverage of tobacco issues, the KDA argues that tobacco no longer matters or perhaps no longer exists in Kentucky. The knowledge that was once so important for the KDA to claim as its own has become a stigmatized knowledge from which the state now attempts to distance itself.

Yet in October 2008 the KDA newsletter—which became an online publication that year—included a link to a University of Kentucky press release on the importance of avoiding quick curing during the dry curing season. Such self-evident coverage of tobacco may continue even as self-conscious attention to tobacco as "heritage" fades. Tobacco may no longer be celebrated by the Kentucky Department of Agriculture, but it continues to be grown by many farmers.

Part 3

Raising Burley Tobacco in a New Century

Introduction to Part 3

When I asked one county extension agent about public perceptions of tobacco today, he replied:

> If you're asking, Ann, if a tobacco farmer in Kentucky can go to a national meeting somewhere, like you know the Community Farm Alliance or Farm Bureau, and stand up and say "I'm a tobacco farmer from Kentucky" and be proud of it—*I* think there has to be a little bit of stigma, that they don't do that. As I've aged in this position and have been able to go to more and experience more national type meetings, or people from other places, yeah, when you—the connotation of tobacco farmer, now, is not that good wholesome, you know producing good food and fiber for the United States. In other words I think yeah there is a—we're—this is a vice.

Over the course of the second half of the twentieth century, the social and political meanings of tobacco have undergone extraordinary changes. As this extension agent pointed out, however, not only have the meanings of *tobacco* changed, but so too has the status of the *tobacco farmer:* there is now a "stigma."

Taken together, parts 1 and 2 of this book demonstrate diverging realities: while tobacco production continues in Kentucky, and many farmers still take pride in what they do, the social and political meanings of the crop have changed dramatically, and tobacco has been relegated to history with the label of "heritage." The final chapters bring ethnographic research and rhetorical analysis together in order to examine two aspects of this divergence as they affect farmers: the complexities of tobacco nostalgia and the disparate meanings of "transition" following the 2004 buyout. Both are deeply shaped by the changed meanings of the crop and the identity category "tobacco farmer."

Farmers raise tobacco in a very different world as a result of these changes, and they acknowledge this in a number of ways, including direct acknowledgment of stigma and mantras of defense. Kathleen Bond, who raised tobacco with her husband and his family most of her married life, told me, "One thing that's really changed is when we got married and people raised tobacco it was a good, honest way to make a living. And at the time that we got out [2003], if you raised tobacco, you were dirt, you know you were contributing to the cancer." Bond contrasted tobacco farming as an "honest way to make a living" with tobacco farmers as "dirt," suggesting Erving Goffman's most basic definition of stigma as "spoiled identity."[1] However, tobacco stigma differs from that considered by Goffman in that the category "tobacco farmer" itself moved from respect to stigma within the lifetimes of today's farmers. Farmers moved with it. Valerie Grigson commented, "It was a prestigious thing to be a tobacco farmer and then. Of course we had all the lawsuits and all that and everybody hates smoking now and, you know, you're a demon if you raise tobacco." The multiple lawsuits against tobacco manufacturers that took place in the 1990s directed a public spotlight on the practices of the tobacco industry and raised questions about who is to blame for tobacco-related illnesses, the smoker or those who make the products—a distinction that is in part dependent on whether the smoker is understood as an addict and therefore a victim or as an agent with free choice.[2] Industry efforts to shift the public focus from tobacco companies to farmers in order to induce empathy and deflect criticism failed to derail anti-tobacco forces. However, these efforts helped to ensure that tobacco farmers were implicated along with tobacco companies. Ann E. Kingsolver notes that "in more polarizing moral arguments about the health effects of smoking and who is responsible for them, tobacco farmers, rather than tobacco companies, have often been portrayed as 'evil.'"[3]

A survey conducted in 1995, as litigation against tobacco companies was heating up nationwide, demonstrates this shared blame. Interviews were conducted with 528 US tobacco farmers and 991 Americans in various regions of the United States who did not grow tobacco in order to ascertain attitudes about and knowledge of tobacco, tobacco farmers, and the federal tobacco program, as well as issues such as diversification and taxes. Of the 991 non-tobacco-farmer respondents, 29 percent agreed or strongly agreed with the statement "Tobacco farmers are responsible for health problems experienced by smokers," and 45 percent agreed or strongly agreed with the statement "Tobacco companies are responsible for health problems."[4] Although more people in this survey blamed tobacco companies than blamed farmers, it is significant that nearly one-third of respondents located blame with farmers.

While the industry rhetoric and the lawsuits can be understood as factors that contributed to generalized perceptions about tobacco farmers, shifting power dynamics contributed to the changed status of growers in their own communities.[5] For much of Kentucky's history, tobacco farmers along with warehousemen and others involved in the industry at the local level wielded a great deal of economic, political, and social power in their communities, as well as in local, state, and national politics. Throughout much of the twentieth century, raising a patch of tobacco, large or small, was the norm rather than the exception. Tobacco farmers were considered a powerful constituency for politicians from tobacco regions, many of whom depended on tobacco income themselves. For instance, a *Lexington Herald-Leader* columnist, writing about legislative issues in the late 1990s, observed: "Tobacco politics in Kentucky used to be easy. To make it with the burley boys, a politician just had to know that blue mold wasn't the stuff on old Swiss cheese, that topping was more than what goes on a pizza, that the companies were out to take what wasn't nailed down and that the tobacco program was next to godliness. You didn't tax tobacco, not a cent. And you talked about 'alternative' crops, about how farmers were too dependent on the leaf, but you never meant it."[6] Kentucky's eminent historian, the late Thomas D. Clark, wrote in his 1977 *Agrarian Kentucky* that "only a reckless politician would suggest support of measures injurious to the Kentucky agrarian way of life."[7] Over the course of Kentucky's history, the needs of the farmer—including and perhaps particularly the tobacco farmer—were central to Kentucky politics, and according to Clark, "Leading their constituencies in and out of crises, these masters have ever paid tribute to the basic precept that the agrarian way was the wholesome way. They promised never to tax it out of existence or to bring it under threat of revolutionary change."[8] According to John van Willigen and Susan C. Eastwood, writing twenty years later, however, "The political influence of tobacco people is decreasing."[9] This is demonstrated, for instance, by recent tobacco tax increases in Kentucky. I asked Alice Baesler, who raised about three hundred acres of tobacco in 2007, how the discussion that took place during the 2008 session of the Kentucky General Assembly about raising the cigarette tax differed from what would have taken place twenty years ago, and her reply is representative of comments I heard from others: "Twenty years ago they would have never even thought it. We were such a tobacco state and tobacco-oriented and all. No—I mean no one would have even brought it up. But of course as the health people and the tides have turned, that's the thing to do—I mean to pick on tobacco because you can pick on tobacco because *nobody likes tobacco.*"

Claiming inclusion in the category "tobacco farmer" has become difficult for some. According to Kevan Evans, a former tobacco farmer who now runs an agritourism business (an orchard and farm stand with events and activities that draw families out to spend time on their farm) with his daughter: "And it got to a point where, you really couldn't go out—you know if you went out of the state and they said, 'What do you do?' 'Well I'm a tobacco farmer'—I mean they kind of frowned on you. You know, but I can go out and say, 'Well I'm a vegetable farmer.' 'Oh okay yeah! Let's talk about it.'" Others continue to proudly claim membership in the category tobacco farmer in ways that, whether with direct acknowledgment or not, suggest defensive responses to stigma.[10] Over the course of my fieldwork, I began to take note of multiple mantras of defense deployed in response to the changing status of tobacco. This includes, as previously discussed, the use of tobacco heritage to remind audiences not only of tobacco's importance in the past but of the ongoing economic importance of this inherited farming tradition. Another common defense is that "people choose to smoke," or "I'm not telling anybody to smoke," echoing industry rhetoric that "smoking is a choice" rather than an addiction. Additional defenses include "tobacco pays my bills" and "it's a legal crop." Like the use of tobacco heritage and smoking as a choice, these can be understood as the redeployment of strategies used by the industry, the state, and even the media. Peter Benson describes similar strategies he encountered among North Carolina growers as "a stock script involving several patterned lines of defense."[11] They may be that, but they are also means by which members of tobacco communities understand and respond to the vast changes they have experienced as not only tobacco but their very identities have moved from respect to stigma during their lifetimes.

"Tobacco Pays My Bills"

It is not uncommon to see truck bumpers in tobacco regions that proclaim, "Tobacco pays my bills." Frequently, I heard lists of all the many things that tobacco has paid for. Such lists are similar to heritage rhetorics, as they serve as a reminder that many Kentuckians have depended for multiple generations on tobacco as a source of income that, according to Jerry Bond, "paid the taxes, paid the insurance, [and] put the kids through college." Valerie Grigson described her dilemma as a mother of children who are taught about the dangers of tobacco at school at the same time that her family depends on it for income: "I'm like how do you tell your kids 'Boys, this is what pays for your house, this is what pays for your car, this is what pays for your Christmas?'" At the same time, Grigson "pray[s] every night they don't smoke." Alvin Bogey told me in 2005:

> I started farming, when I was about six years old, that's when my dad bought a farm and I drove a team for him then. And we raised tobacco every year. Tobacco has meant the livelihood of my father and the family. It has meant the livelihood of myself. If it hadn't been for tobacco, my two daughters wouldn't have had a college education. I put them through college and they both taught school and have retired now. Tobacco has made that. Tobacco has got some bad points, but it's got some good points. If it hadn't been for tobacco, there'd [have] been some kids wouldn't have shoes in the winter time, wouldn't have got to go to school, or maybe had some food on the table.

This mantra also argues that poundage prices have not kept up with farm expenses to the extent that tobacco income can no longer do all of the things named in these rhythmic lists, including paying for a farm. Tobacco once meant farm stability, but there is no longer much of anything that is stable about the family farm (and indeed this was from the beginning a tumultuous industry). And yet "tobacco pays my bills" argues that tobacco remains economically important, in the present tense, to many.

"It's a Legal Crop"

Another steady refrain running through my fieldwork were forms of the statement "It's a legal crop." County extension agent Mike

Carter told me, "A lot of growers will tell you that 'It is a legal crop. It is a potentially profitable crop. There are companies that want to buy it, and are willing to pay me a fair price for it. As long as it's legal, and I can make money doing it, I'm gonna continue to grow it.'"

Extension agents often talked about the legal status of the crop, perhaps reflecting pressure from constituents who oppose the university's continued work with tobacco. In a discussion about the growing lack of knowledge in the general public about agriculture, Lincoln County extension agent Dan Grigson and I had this exchange, which suggests that he may be responding to criticisms he has experienced:

> Author: Is it harder—or would it be harder to educate people about, tobacco versus food—
>
> Grigson: Oh yes. It's much much harder. Because so many folks are anti-tobacco because of the health issues that are related to it. There are many folks who would say "Well I don't want to hear anything about tobacco, you ought to totally quit that anyway."
>
> Author: And Kentucky folks?
>
> Grigson: Yes, even Kentucky folks yeah. I mean but it's still a legal product and as long as it is a crop that we can grow I think we need to continue to grow it, we need to continue to educate folks about it.

The legal status of tobacco was raised by farmers as well, often in direct statements that "as long as it's legal and I can make a profit, I will keep raising it." The refrain "it's a legal crop" vividly suggests that farmers feel as though they have been treated as criminals although they have not broken the law. Growers also told me stories about being compared with growers of illegal crops such as marijuana and opium poppies, comparisons that they found deeply offensive. One grower told me such a story when I asked about the license plate on the front of his pickup truck that read, "Tobacco pays my bills," linking these two defenses.

While for outsiders the line between immoral and illegal may be gray (or nonexistent) in regard to raising a harmful crop, there is a clear line for many tobacco farmers. One farmer told me, "If they make it illegal I'll quit raising it"; another told me a story of someone trying to convince him to grow marijuana on his land and of his

Marlon Waits, along with other farmers, proclaims "Tobacco Pays My Bills" on his license plate. (Photo by author)

refusal. Such instances demonstrate the importance to many of following the law, even as others judge their morality based on the crop they raise.[12] The frequent assertion that "it's a legal crop" is also an attempt to side-step the issue of whether or not it is "moral" to grow the crop, either because farmers themselves question the morality of growing an addictive, lethal crop or because they know that this is a battle they cannot win. Tobacco's morality has been impeached. They therefore use the unimpeached argument that it is "legal."

At the same time, comparisons to alcohol serve as reminders of the fluidity of morality. When I asked Noel Wise if he felt like the media made "judgments about tobacco farmers," he responded with this story:

> I grew up—like most other people I grew up going to Sunday school and church every Sunday. When I was a youngster I remember, particularly one elder in the church where I went to church that, they were debating about him being an elder because he was employed part-time at a distillery. And they made alcoholic beverages for sale and people got drunk and that was a terrible sin. And the tobacco farmers were about to kick this here distillery worker out of the church. I think that's turned around now, now you can drink that alcohol with no ill effects if you don't overdo it. But now the smoking's a sin.

Through this narrative, a response to my question about feeling judged, Wise described a fairly common perception that alcohol and tobacco have switched places in terms of social acceptance, and therefore so too have the producers of these substances.

Each of these defenses acknowledges the changed status of tobacco while also defending ongoing production. These are just a few of the many ways farmers and others express their understanding of the changed status of tobacco and tobacco farming, and they provide important context for the remaining chapters. In chapter 6, I consider the combined impact of the technological changes and the changing political status of the crop on the traditional "pride" farmers take in their crop. I examine particular expressions of tobacco nostalgia that communicate feelings of loss for a better time of tobacco production next to the idea that, for some, "now is the good old days" because of technological innovations and improved efficiency. Through such discourses, tobacco farmers express a longing not for a return to earlier times and technologies—which would be economically unfeasible—but for the pride and respect once associated with a "tobacco-man" identity.

In chapter 7, I examine competing discourses of "transition" as they relate to ongoing discussions about the future of agriculture in Kentucky. The dominant perspective on the future of Kentucky agriculture is that tobacco production is in its last days and that the "transition" period currently taking place is one in which tobacco farmers have replaced, will replace, or should be replacing tobacco production in favor of "diversified" agriculture. This rhetoric suggests that simply raising alternative crops and/or livestock will lead to the replacement—economically as well as symbolically—of tobacco. In contrast, many farmers who are choosing to continue to raise tobacco do so based on the belief either that there will always be a market for Kentucky burley or at least that they will raise it until there is not. For them "transition" refers to the end of the federal tobacco program and the negotiation of new relationships with tobacco companies through direct contracting. This period of transition also challenges us to examine the multiple meanings and uses of the word *tradition* in the context of both diversified and tobacco farming.

Chapter 6

"*Now* is the good old days"

Burley Tobacco Production and Nostalgia

Living in Kentucky, I came to understand that most people have family ties to tobacco, and many worked in tobacco as children or young adults. This is not surprising; I was repeatedly told that at one time "every little farm" had a tobacco base, and indeed, throughout the twentieth century tobacco was grown on most Kentucky farms. People talk about tobacco work with a mixture of emotions, and descriptions from memory are often saturated with sensory details. They describe the comforting smell and atmosphere of the tobacco stripping room on winter days but also the oppressive heat and stickiness of working in the fields, suckering, topping, or cutting tobacco. But even the memories of miserable work are often framed in prideful reminiscences of having worked hard and having participated in an experience shared by so many Kentuckians. Tobacco is woven tightly into Kentucky culture, and many feel that tobacco made Kentucky what it is today.

Many also lament lost family time, as raising the crop once depended on the entire family working together during particular stages of production. Tobacco is often referred to as "the glue that held families together." Not only did tobacco mean time together as a family, but tobacco was the economic mainstay—it "paid the taxes, paid the insurance, [and] put the kids through college," as Jerry Bond told me in 2005. The crop was, at one time, the sole source of cash for many families, and I was told over and over that there was a time when the value of a tobacco base was enough to serve as collateral on a farm mortgage because it meant a guaranteed income. Tobacco quite literally made it possible to own a farm, for full-time farmers as well as for those who worked off the farm and either raised their small tobacco quota or rented it out to others. Tobacco also helped Kentucky maintain a small-farm culture far longer than other regions of the country.

The objects of tobacco nostalgia are many and are dependent on the context and on the speaker. Some expressions of tobacco nostalgia resemble the more general nostalgia many feel for a rural life of the past, which, from the perspective of the present, is described as a simpler, happier time. Other expressions of nostalgia are quite specific and unique to tobacco. One such thread of nostalgia focuses on the production of the crop itself, and that is the subject of this chapter. In 2005, a man who has worked in tobacco part-time all of his life, but who does not raise his own crop, told me that today "they treat it like . . . baling hay or something." He went on to say: "They just pick it up by handfuls sometimes and throw it in there. And when you've worked the other way you just can't, you don't treat tobacco that way." This comment speaks volumes about the reverence with which the crop has historically been treated within tobacco communities. Raising tobacco is often described as "an art" because of the specialized knowledge required at every stage of production—knowledge that is sometimes described as residing in the blood rather than as learned. Because so much of burley tobacco production continues to be carried out by hand, burley has long been understood as a handcrafted farm commodity. I assumed that the perspective of the tobacco worker quoted above was representative of what I would continue to hear as I interviewed more and more growers—that farmers were moving away from traditional and "right" ways of doing things and no longer treated the crop with the respect it deserved, with pride. I have heard additional, similar lamentations, but expressions of nostalgia are far more varied and complex than I initially understood. They are dependent in part on the position of the speaker in the present.

In his examination of the expression of nostalgia through the collection of material culture by members of a Northern Ireland community, Ray Cashman demonstrates that the immediate object, the literal thing, that nostalgia appears to be about is in fact symbolic of much more. He argues that "nostalgia is a cultural practice that enables people to generate meaning in the present through selective visions of the past."[1] Similarly, I argue that expressions of nostalgia for the ways in which tobacco was once produced are a stand-in for more complex losses, particularly the changed status of *tobacco men*. Like the nostalgia Cashman describes, the nostalgia expressed by tobacco farmers is not about a desire to return to the past, but a comparison of past and present that highlights what has been lost as well as what has been gained.

Mourning a Lost Way of Life

In 2004, Kentucky photographer James Baker Hall and poet and author Wendell Berry published *Tobacco Harvest: An Elegy*, a book of photographs taken by Hall during a 1973 tobacco harvest that took place on a farm neighboring Berry's farm and in which Berry and members of his family participated. The photographs together with the accompanying essay are to be read, according to the title, as a mournful poem about the end of Kentucky's tobacco farming tradition. It is an end that Berry argues began in the years immediately following this harvest. He writes: "In 1973 James Baker Hall photographed, with acute discernment, these scenes and events of a Kentucky tobacco harvest. The place, the work, and the people are mirrored in these pictures as they were. We look at them now with a sort of wonder, and with some regret, realizing that while our work was going on, powerful forces were at play that would change the scene and make 'history' of those lived days."[2] For Berry, the Kentucky tobacco harvest is "history" viewed by the participants "with a sort of wonder, and with some regret" that "the place, the work, and the people" in the "scene" have changed. Berry's use of the phrase to "make 'history'" serves to disconnect tobacco harvests of the past from those of the present and to suggest that the cultural practices in the photographs exist no more.

Actually, the work involved in a burley tobacco harvest has not dramatically changed. As for place, tobacco is still being raised on many farms in Berry's home county, Henry. According to the 2007 Census of Agriculture, Henry County ranks seventh out of the 106 Kentucky counties (of a total of 120 counties) in which tobacco is produced.[3] The people who make up such a scene, however, have in most cases changed—and here is the source of Berry's nostalgia: he mourns the end of the coming together of neighboring farmers to harvest a crop of tobacco. As I have described, migrant workers, primarily from Mexico and primarily undocumented, have become the principal source of labor during cutting and housing season for most Kentucky farmers, as they have for farmers of other crops across the country, now accounting for at least 75 percent of total tobacco labor hours.[4] Berry regrets the dependence on a "subordinated class of menial laborers working without either a proprietary interest in the crop or equity in the land."[5]

The incongruity in Berry's regret, however, is that there had long been a reliance on hired labor during the tobacco harvest season and at other times as well, and even in the period represented in the photographs the labor situation involved inequities based not only on race but also on gender and class. As Ann E. Kingsolver argues, tobacco is often glossed "as a family-based cash crop," and those who have been paid to work in it have often been ignored.[6] In many counties, African American men and women worked in tobacco for wages or as tenant farmers (and also as owners of farms). Low-income widowed women worked in tobacco, particularly during fall tobacco stripping rather than during the harvest, although some did help then too. People in the counties further to the east talk about the "boys from the mountains" who would come at harvest time, boys from the Appalachian region for whom the harvest was perhaps an annual opportunity to earn cash. And every county had sharecroppers of varying economic means.

In my fieldwork, hired labor often went unmentioned at all until I asked about it. I was also frequently told that until the most recent decades, labor was hired almost exclusively for cutting and housing and sometimes for stripping tobacco. In my examination of the Kentucky Department of Agriculture newsletters, however, I noted periodic references to hired help in tobacco, as well as labor shortages, in seeming contradiction to the widespread nostalgia for a time in which tobacco work was all done by family labor. As just one example, in 1954 there was a description of what was then a new chemical being tested, maleic hydrazide, which "in the very near future" "may eliminate hand suckering" of tobacco plants.[7] The KDA estimated that maleic hydrazide—now known as *MH, sucker control,* or *sucker dope* and used universally in the region in place of hand-suckering—would cost the farmer about fifteen dollars per acre but would eliminate about thirty dollars per acre in labor costs. The fact that labor savings, quantified in dollars rather than time, provided a rationale for the use of MH suggests that the hiring of labor for suckering was not unheard of.

Berry has carved out a particular time and place to mourn, characterizing the "historical moment" captured in the photographs as an "interlude" between periods in which tobacco's symbolism had changed and the people doing the work had changed.[8] This act of carving out a particular moment as representative of tobacco history recalls John Dorst's reading of Chadds Ford, Pennsylvania, in which he argues that the Chadds Ford Historical Society places the "definitive past" of the region in the colonial period and in doing so "reduces the synchronic richness of a past period, with all its complex social forces and relations, to an immediately comprehensible ensemble of a few elements—a kind of tableau."[9] According to Stuart Tannock, rhetorics of nostalgia always involve "a positing of discontinuity" between "then" and "now."[10] Berry has created a tableau in which the complex history of tobacco work is simplified and symbolically reduced to one harvest, a harvest of "then" that is discontinuous from those of "now." As an "interlude," this moment is discontinuous with tobacco harvests as they still take place, for while Berry mourns the death of the pride with which the crop was treated, many farmers continue to take pride in their tobacco crop.

Not only has Berry simplified the labor situation of the early 1970s and suggested that today's farmers no longer demonstrate pride in their crops, but in his tableau he glosses over the changing politics of tobacco, marking this time with the statement that, looking back, "we" realize that "powerful forces were at play that would change the scene and make 'history' of those lived days."[11] Here he implies that he and the others in the photos were unaware that the changes going on around them would affect them. Yet, among other things, tobacco farmers had just experienced the release of the surgeon general's report on the dangers of smoking in 1964, the imposition of a warning label on cigarette packages beginning in 1965, the removal of cigarette advertising from the airways in 1971, and the continued rise of overseas tobacco production. Change was clearly in the air, and the awareness of it is most likely what prompted the taking of the photos.

Berry states, "I no longer remember how these pictures came to be made."[12] Taking family photographs in the tobacco crop each year has long been a common form of vernacular documentation.

According to extension agent Keenan Bishop, "You look through the family albums and you see, you know, pictures of kids' birthdays and Christmases and tobacco crops." However, the photographs in this book are contextually quite different because James Baker Hall was an artist, not a family member or participant in the harvest.[13] Hall was documenting the harvest, not taking snapshots for the family album. It is important that during this period, as previously described, tobacco as "a way of life" was becoming art, through the release of "heritage prints" and other artistic renderings of tobacco work. That the photographs were made at this time suggests that they resulted from growing self-consciousness of tobacco as threatened heritage.

This points to important connections between heritage and nostalgia, as well as distinctions that are not often articulated. Heritage refers to inheritances, both tangible and intangible, that are perceived as rights granted by birth, such as folklore, culture, values, belief systems, history, ways of living, possessions, and land—but it is a term that is most often applied when such things are understood as threatened. When the photographs were taken, they may well have been intended to document threatened tobacco heritage. However, upon publication thirty years later, the volume is a production of nostalgia rather than heritage because of the perspective within which Berry frames it: one of total, painful loss. Berry's essay can be understood as his expression of mourning for his own heritage.

Tobacco Harvest: An Elegy is also a classic expression of nostalgia because its major message is that in a comparison of how tobacco is raised today with how it was once raised, today's tobacco production comes up short. According to Fred Davis, nostalgia "is a past imbued with special qualities, which, moreover, acquires its significance from the particular way we juxtapose it to certain features of our present lives."[14] The differences Berry alludes to are also often pointed out by others. While some small farmers still touch every leaf of tobacco that they grow in processes identical to those pictured in Hall and Berry's *Elegy,* others don't touch a single leaf but instead supervise the work of others. Although there may have always been hired labor, the tobacco man and his family worked beside them. This is no longer the case for many farmers, resulting in a changed relationship with the crop. This changed relationship, and perceptions that the crop is differently treated as a result, is central to tobacco nostalgia.

The Golden Age of the Tobacco Man

Until recent decades, tobacco production was dependent on the work of the entire family, with or without hired labor, and often each family member had specific roles to play at each stage of the process, based on gender, age, and ability. Divisions of labor varied by farm size and economic position. Generally speaking, women traditionally participated in tasks that required multiple people, such as pulling plants, setting, topping, stripping, and occasionally cutting and housing, and they cooked large meals for everyone working in the tobacco, including family, neighbors, and hired labor. This varied by social class, however, as in some cases women had no involvement in tobacco work at all.[15]

Whether women worked in tobacco or not, the crop and the work have most often been controlled and monitored by men, and men have traditionally done most of the cutting and housing, as well as tasks that required driving a tractor or other farm equipment. Tobacco is certainly not unique in this regard. The work on American family farms more generally has traditionally been directed by men and carried out by men, along with women at times when additional labor is needed for particular tasks.[16] As times have changed in tobacco—as children have moved off the farm, women have gone to off-farm jobs, and acreages have grown—women's roles have in many cases decreased even further or ceased altogether. Men have had the primary multigenerational relationship with tobacco as a crop, a craft, and a source of occupational identity, and with important exceptions—particular farms on which women are heavily involved in tobacco production—it is men who are most involved in continuing the tradition today. This relationship is often summed up in the use of the term *tobacco man*.[17]

At one level, "tobacco man" references all men involved in tobacco production. Over the years, the term was used fairly regularly in the Kentucky Department of Agriculture newsletter, in such contexts as "One of the most discussed questions among the tobacco men is, 'What to do with respect to allotted acres.'"[18] In his 1947 narrative of life and work as a Kentucky tobacco farmer, Virgil Steed noted that "local farm tenants call themselves 'tobacco-men.'"[19] However, the term does not simply differentiate between farmers who grow tobacco and farmers who grow, say, corn or soybeans. Rather, the term implies characteristics and concerns that distinguish a "tobacco man" from other categories of men. Wendell Berry described the social meaning of the category this way: "As a boy and a young man, I worked with men who were as fiercely insistent on the ways and standards of their discipline as artists—which is what they were. In those days, to be recognized as a 'tobacco man' was to be accorded an honor such as other cultures bestowed on the finest hunters or warriors or poets. The accolade 'He's a *tobacco* man!' would be accompanied by a shake of the head to indicate that such surpassing excellence was, finally, a mystery; there was more to it than met the eye."[20]

"Tobacco man" symbolizes a gendered relationship with the crop that encompasses the performance of particular skills, knowledge of an aesthetic system, and a type of man who is highly respected in the community. The days of the tobacco man are widely perceived as going or gone, as there are fewer and fewer tobacco farmers, and as many of those who remain have changed their farming practices. As tobacco became stigmatized, the status of a category of men who were once among the most respected in the community also changed.

County extension agent Dan Grigson described to me the lessons he learned from his father about the importance of doing a job the right way at the right time, particularly after he'd been given two-tenths of an acre of his own tobacco to raise. I asked if he would describe the farmers whom he now works with daily in the same terms that he described his father, and he replied:

> Sure there are a lot of good tobacco farmers. There are folks who grow tobacco. There are folks who I call, "pretty good tobacco farmers." And then there's that, that upper level who just they're "tobacco men." They're tobacco women or tobacco men. But they do the extra, they always seem to have a good crop, even in a dry year or, you know, too wet of a year, they'll still come out with a very good crop of tobacco. Detail people. People again who are good managers and make things work. They take care of getting the soil samples and making sure the fertility's right. Selecting the best varieties, making sure pests are under control. Topping at the right time, cutting at the right time, and watching those doors and curing that crop down. And then, take a lot of pride in making sure that when they prepare it for market it's prepared well. Not just all thrown together and, grades are not mixed up,

> no weed trash or anything like that in there. They truly have pride in doing a good job.[21]

Grigson articulates prevalent ideas about what makes a good tobacco man. A tobacco man is someone who knows the proper cultural practices required to raise a good crop and carries them out consistently because he or she has "pride." Tobacco men are farmers who continue to treat tobacco with respect and pride.

Particular farmers I met over the course of my research were clearly admired by others because of their continued prideful performances of traditional practices. Many of the tobacco men I got to know were smaller farmers who persist in doing much of the work themselves, beside those they hire, and who are viewed as continuing to take pride in quality as well as quantity. Such farmers are regarded as no longer typical, and the prevailing opinion is that their days are numbered. In his study of Oklahoma farmers, Eric Ramírez-Ferrero suggests that smaller, "family-focused" farmers were "considered noble yet naïve and anachronistic." Such farmers, he argues, "were being increasingly marginalized as industrial discourses gained prominence in the region."[22] Small tobacco farmers, too, seem to be regarded as "noble but naïve," and I was often told that farmers now face a choice based on the economy of scale: they can either raise many acres and get paid less per pound because of poorer quality or stay small, do the work themselves, and get paid more per pound for a smaller amount. But there is a widespread consensus that the days of the small tobacco farmer are numbered in large part because the changes coming down the pike—from big bales to mechanical harvesting technologies—will be out of their reach. Berry's *Elegy* is a public expression of a nostalgia felt much more widely for the tobacco man and for a past Golden Age of tobacco production in which tobacco men were among the most respected members of the community. Such men touched every leaf of tobacco multiple times because of the pride they took in their crop.

"Golden Age" refers to a period in time that is widely understood to have been somehow better and therefore a period about which nostalgia is expressed. Timothy Lloyd and Patrick Mullen describe a "Golden Age" of commercial fishing, illustrated in the personal experience narratives of retired Lake Erie fisherman. In these narratives, the men describe the past as always a better place because there was "more primitive technology, harder work, clearer water, more abundant fish, and less governmental restriction."[23] According to Lloyd and Mullen, "The work was hard, but the men were up to it. This is a significant element in maintaining occupational identity after retirement: the retired fisherman can still see himself as a fisherman because from his perspective the ones fishing today are not fishermen in the same sense that he was."[24] Similarly, Amy Shuman has described a "Golden Age" period for a community of artisans in Pietrasanta, Italy, in which "stories serve as nostalgic reminiscences that glorify a lost past and lament the present state of events."[25] Such stories act to construct and maintain a shared understanding of the period of these stone carvers' grandfathers as a Golden Age, before particular changes occurred. These artisans "are constructing this scenario of disappearance by determining what counts as continuation and deserves to be lamented in change."[26] The narratives they tell, according to Shuman, are allegorical stories of the past that comment on the present.

These two examples highlight the role of "change" in the narrative construction of a Golden Age. Berry's delineation of 1973 as a Golden Age, while determined in part by what he presents as the happenstance of Hall's photographs having been made that year, highlights a period in time prior to the most commented-upon changes in the context and production of tobacco. This is highlighted by Berry's comment that "the Henry County tobacco patch of 1973 was not remarkably different from that of, say, 1940."[27] Of course it was different in many ways, but Berry looks back at it and sees similarity rather than difference because of the scale of the changes that have since taken place. This harvest took place between the release of the surgeon general's report and the point at which farmers fully felt its effects. This harvest immediately preceded the technological change that is most often cited as the biggest, the move from hands to bales, and it was the end of the period when acreages were small enough that a group of family and neighbors could get much of the work done and local labor could be depended on when needed. And this harvest took place before "tobacco farmer" became a stigmatized category. This combination of changes is important because the social and political meanings of the crop were changing at the same time that farm and labor practices were introduced that changed farmers' relationship with the crop.

For these reasons, it would make sense for the period that Berry is eulogizing to be thought of more widely as a "Golden Age." However, on those occasions when I specifically asked interviewees the question "What was the best time for tobacco?" I was most often told it was the 1980s or '90s. This suggests that there are either multiple Golden Ages, that there was no Golden Age at all, or that my use of the term "best time" had a meaning that is different from what might be understood as a "Golden Age." In some ways, these are simultaneous truths. The 1980s and '90s were the best times for farmers because tobacco was in demand, leading to better prices and the ability to strip it into fewer grades, which meant labor savings. The recent move to baling tobacco had also resulted in enormous labor savings, and quota lease arrangements improved during the period. These were the "best times" when measured by economic prosperity, but they are not the times about which nostalgia is most often expressed.

During a 2008 interview with Martin Henson, he and I discussed the current push for farmers to move to large bales as compared to the circumstances of the move from hands to bales. He commented that "things have really changed" in regard to the great reduction in the amount of time it takes to strip tobacco as a result of the move to baling, and we then had this exchange:

> Martin: But oh we handled it a whole lot different, whole lot different. Everything had to be so-so.
> Author: And do you miss that?
> Martin: No [*laughing*]. We are in the good old days now. Just like several years ago I had a, old gentlemen that helped me for years. And he was up about, eighty years old at the time. We were stripping tobacco. Harold was with us. And we got to talking about the good old days. Me and Harold was talking about the good old days. [The older gentleman] after a while he got tired of hearing that. He said "I wanna tell you boys," he said, "right now is the good old days" [*laughter*]. We was talking about the horse—you know, using the horses. Of course when I first started out, that's what I used, I used horses to plow my ground with, cultivated with, and the whole works. Had a tobacco setter, pulled it with a team of horses. That's the way I started out.
> Author: But those weren't the good old days.
> Martin: Those weren't the good old days [*laughter*]. *Now* is the good old days.

I think Martin was laughing at his own nostalgia here, knowing that he wouldn't anymore want to go back to tying hands than he would want to return to using a team of horses to plow and cultivate. His laughter is an acknowledgment of his awareness that he and other farmers do feel and express such nostalgia for earlier times, even though they would not choose to go back to such times. Some things were better in those days, and other things are better today.

This instance of nostalgia supports Ray Cashman's persuasive argument that "not all nostalgias are the same."[28] According to Cashman, scholars still understand nostalgia in the context of its meaning when it was coined—"from the Greek *nostos,* 'to return home,' and *algos,* 'a painful condition'"—to describe a "potentially fatal form of homesickness" that came to be understood as "a disturbed frame of mind that distorts the past in response to personal, irrational desire."[29] Farmers' expressions of nostalgia for past times, coupled with strong statements that they'd never go back, are *rational* articulations of the belief that there were things that were better in those days *and* there are things that are better today. Like Cashman's collectors and community historians, tobacco farmers "quite sanely challenge both the presumption that modernization equals positive progress and the impulse to romanticize the past."[30]

Throughout my interviews and conversations, nostalgic stories and expressions of change arose that are comparable to Lloyd and Mullen's and Shuman's examples of Golden Age narratives. The period(s) understood to be the good old days, or a Golden Age, differ by speaker, however. Rather than referencing a consistent time period, nostalgia was most often expressed about the tobacco raised by the speaker's father or grandfather and the farming practices of these earlier generations of men. One day, during my fieldwork, a twenty-year-old tobacco farmer asked me if I had a title for my book yet. I told him I did not yet have a title and asked him what he would call it, and after some prodding, he finally said, "Oh I don't know, 'The Last of a Dying Breed'?" I was surprised at this suggestion because he had previously expressed nothing but confidence about the future of tobacco, and he had repeatedly told me of his hopes of increasing, even doubling, his family's already large tobacco acreage. However, I came to understand that this title suggests his perception of difference between his generation and the "breed" of tobacco farmers who came before. I believe he meant that "tobacco men" are a dying breed. Fred Davis argues that "the past which is the object of nostalgia must in some fashion be a personally experienced past rather than one drawn solely, for example, from chronicles, almanacs, history books, memorial tablets, or, for that matter, legend."[31] Examples such as this suggest that this is not always the case. This young man was nostalgic not for a personally experienced time or place, but for a category of men whom he sees fading away.

Cashman describes the use of nostalgia in the Upper Derg Valley of Northern Ireland as a critique of the changes that have taken place over the years and argues that in this community nostalgia is a way of coping with accelerated technological change. Although tobacco nostalgia, too, can be understood as a means of coping with change, farmers often celebrate recent technological changes and innovations. The following exchange took place during an interview with Martin, for instance:

> Martin: This technology's great [*laughing*]! It is.
> Author: Sounds like it.
> Martin: Things have come a long way, come a long way. I've often wondered what my daddy would say if he saw me baling tobacco [*laughing*].
> Author: What would he say?
> Martin: I don't know [*laughing*]. He's probably turned over, several times.
> Author: Why?
> Martin: Well, I know, for years he raised tobacco, he raised tobacco until he was eighty-four when he passed away. He raised tobacco for probably seventy some years, seventy-five years. [He] hand tied it, everything was neat, just, prim and proper. Now you just throw it in there and tramp her down and go on [*laughing*].

This exchange followed Martin's description of the move from plant beds to float beds, the second most commonly cited technological innovation in recent tobacco production history. The move to bales is often described as *the* biggest change that has occurred in tobacco production since the introduction of maleic hydrazide or sucker control in the early 1960s. The parallel between the move to bales and the move to float beds is emphasized by Martin's shift from one topic to the next, connecting them with his exclamation "This technology's great!" He then immediately, however, shifts to his father's perspective, a common occurrence when innovations such as baled tobacco are mentioned. Speakers often followed positive comments about innovations with a coda similar in meaning to Martin's statement—*what would my daddy or granddaddy say if they could see us now?* The answer almost always suggests that the previous generation, now deceased, would not approve.

While Martin Henson described his father "turning over" in his grave at the idea of baling tobacco rather than carefully tying it into hands, Clarence Gallagher, a generation younger than Martin, described the neatness of his own father's bales with awe. Clarence described a particularly messy load of tobacco that he had seen, and I asked him what his father would have said about it. He told me he wished he had pictures of his father's bales so that he could show me "what older people did" and went on to say that his bales looked like "you could just shoot a rifle right down the side end of them you know what I mean?" He told me that "everything was just neat." The fact that these speakers, and therefore their fathers, are of different generations suggests that the Golden Age of tobacco production is fluid and based on the expectations and practices of the previous generation. This is in marked contrast to Shuman's description of the Golden Age for Italian stone carvers, which was quite specific: "a period of the early 1900s, when artisans worked in large studios employing hundreds of workers."[32] Martin's description of the current treatment of tobacco—"Now you just throw it in there and tramp her down and go on"—is similar to that of the county agent quoted above who said that a tobacco man's tobacco was "not just all thrown together." However, Martin's judgment on the treatment of tobacco is framed by what his father would think. Because today's farmers learned to raise tobacco from their fathers—whom they remember as true tobacco men—they judge themselves based on their perceptions of the standards of their fathers' era of tobacco production.

Those who do not raise tobacco in the present often shake their heads at the way it is now raised, offering judgments that are similar to farmers' perceptions of what their fathers would say. I asked another county agent, one who had worked in tobacco as a young man, if there was still the same pride in the crop as there once had been. I asked, "How about the pride—you talked earlier, you've mentioned several times the pride in your crop and, I mean is that still there?" and the agent responded:

> No. Well. I mean obviously everybody wants, a good crop—but, they're just worried about getting the pounds and having the

> color that the company wants or the texture that the company wants. You know they're not—they don't care about the weeds in it as long as there's not so many weeds that they can't get the cutters through. You know they don't care about stepping on the tips and bruising the plants. They care about stalks splitting out and falling off because that's, that's a whole stalk that doesn't make it to the barn. But, not you know the leaves. The pride's not there like it used to be. They—there's obviously varying degrees of, there's those [who] still do take a lot of pride in how things look and how well they manage the crop, then there's all the way to those that, you know they don't care, as long as they get a paycheck—and you know those, that extreme are the guys that, you know that, throw trash in the middle of the bale and cover it up with tobacco and, hose the tobacco down with water so it weighs a lot before it goes to the warehouse and stuff like that.

One retired tobacco farmer told me, "For people that raise it now, it's all about production and money. You know they put out large acreages, they don't care if it gets wet." He went on to say that "they just don't produce the product like I was taught to do," which meant following a philosophy of "if you're gonna do it, do it right." Former farmers and workers, like Lloyd and Mullen's retired fishermen, view themselves as having done things in a different, better way. The judgments of those who no longer raise tobacco are similar to perceptions of the views of a past generation. In the meantime, farmers themselves must take the changed economic circumstances of raising tobacco into account as they judge themselves.

The combination of increasing economies of scale, hired labor, and technological innovations has led to changed practices and therefore to judgments that current practices conflict with what it once meant to be a good tobacco man. Lost practices include waiting for just the right weather before cutting tobacco so that it didn't get rained on in the field, managing the curing process by opening and closing the barn's side vents and doors, grading tobacco in multiple grades, carefully collecting fallen leaves, and so on. As acreages have grown and more labor has been hired in, such practices have become impossible for most farmers to continue, yet farmers feel the eyes of their fathers looking over their shoulders.

Lloyd and Mullen note that commercial fisherman "spoke of how hard the work was in the old days compared to today."[33] Retired tobacco farmers do so as well, but just as frequently younger men are the ones making the comparisons, judging their lives as easier and themselves as lazier. Clarence Gallagher told me, "We're lazier. I raise more tobacco than my dad did, but the man worked harder. He worked harder. He worked harder than I ever worked, as far as raising tobacco. He sprayed it all by hand, I've got a highboy. I can spray in one morning what it would take him, two days to spray."

At one time it would have been easier to describe a typical burley tobacco farmer and the work he and his family engaged in because acreages were closer in size. Larger acreages, as well as a much larger range between the smallest and largest farmers, have led to less typicality in terms of farmer involvement with the crop. While many farmers remain physically involved in working ground, setting part or all of their tobacco, and cultivating their crop, they may or may not top, cut, house, or strip any of it. The aspects of the work that today's larger growers are most directly involved in are those that can be done from the tractor or truck, while they are least involved in those activities that involve physical contact with the plant. The loss of physical contact is representative of the changing relationship with the crop.

My first experience stripping tobacco is illustrative of this. I was in a nonconventional stripping room on a large tobacco farm and thought that I recognized the differences between this and more traditional tobacco stripping arrangements. But because this was my first time stripping tobacco, one detail did not stand out to me as nonconventional: in this stripping room everyone wore cotton gloves while they stripped the tobacco. The young farmer I was accompanying loaned me a pair of gloves for the short time I stripped tobacco with him and the others. Later that day, I bought a similar pair of gloves in order to be ready for my next stripping room experience.

I never wore my new gloves. As I visited other stripping rooms, and stripped tobacco myself in some of them, I watched for gloves. The few places I saw gloves worn were stripping rooms in which Latino workers were doing all the stripping, with little or no participation by the grower. When farmers still did some or all of their own stripping—and these were the instances when I stripped tobacco, with them—gloves were not worn, except perhaps when piles of tobacco were being brought in or stalks brought out. If anyone was wearing gloves, it was the hired labor, not the growers.

The gloves are as important symbolically as they are tangibly.

Gloves protect the wearer from the sticky residue of the cured tobacco and keep hands warm in cold stripping rooms. But tobacco men don't need protection from the crop or the cold. Gloves also block the wearer from full access to the plant, inhibit natural movement, and lead to rougher treatment of the leaves; gloves lessen the ability to handle the crop gently (with respect). Not only do gloves provide a literal and symbolic separation from the crop and therefore tangible evidence of the changed relationship with the crop, but they also symbolize the move to tobacco production as "business," as it is primarily hired hands that are protected by gloves.

Comments that tobacco production is "a business" today are frequently used to differentiate tobacco in the present from tobacco farming in the past. In an interview for the Kentucky Oral History Commission in 2001, Christian County tobacco farmer Bruce Cline compared his operation with a neighboring farm, a father and son who raised only ten or twelve acres and "do *every* bit of it themselves." He described these farmers as "very content" and "probably as happy a lot of people as I know of anywhere." According to Cline, "The wives help them strip the tobacco and it's very much a family, oriented business. My farm on the other hand, is a much different beast. It is you know it is more a *Wal-Mart* so to speak." Mark Roberts told me:

> For lack of a better word, I wanna say it's not fun anymore? And I know work is not supposed to be really fun anyway? But it used to be, years ago, family, friends would work together, everybody would jump in and, you would help each other out. It was kind of fun to go to the field and work. Now pretty much, people that raise very much tobacco at all, it's nothing but a business. That's all it is. And it has nothing to do with enjoyment of the crop. It's hard labor, it's all management. There's no time. The farmer that actually owns the land, that's raising a big crop, doesn't have time to get out there and do the labor anymore.

When Mark says it is a business and "that's all it is," he implies that it was once something more. "It's nothing but a business" now is a summarizing phrase of all the changes that have taken place as tobacco acreages have grown, the entire family is no longer involved, and dependence on hired labor has increased. Cline and Roberts both imply that a "family oriented" farm is a happier farm than a farm large enough to be comparable to Wal-Mart. In his discussion of the changes on North Carolina tobacco farms resulting from the end of the tobacco program and the introduction of contract production, Peter Benson echoes Cline's sentiment that farmers work in a new environment in which they have fewer opportunities for taking pride in their work: "Local cultural values of independence and pride are challenged as farmers are reclassed as service providers who are part of corporate teams."[34] Like "pride," "business" serves to describe the cultural practices of tobacco farmers; unlike "pride," however, "business" can be used with contrasting implications. It can be used as either a negative judgment or an acceptance of new practices, depending on the speaker.

"Businesses" and "family farms" are presented discursively as separate categories that imply different relationships with the crop, as well as different lifestyles. For some, "business" is an objectionable word that reflects the speaker's nostalgia for a time when tobacco farms were "family farms," not "businesses." At times, this perspective seems to ignore the fact that tobacco production was always about earning an income. This sentiment was evident in the perspective of the county agent quoted above who said that "they're just worried about getting the pounds and having the color that the company wants." Farmers were always worried about producing as many pounds as they were allowed by law and providing the types of tobacco that the companies wanted in order to maximize their income. However, like Mark Roberts, this county agent was arguing that that is not *all* that they were worried about. Tobacco has intertwined values, economic and symbolic. The tobacco-man pride that is associated with "family" farms is threatened, as for most farmers the ability to touch every leaf and therefore produce a crop to be proud of is gone. So, too, are the opportunities to demonstrate this pride to peers—other farmers—at the tobacco warehouse, as farmers now drop their tobacco off at receiving stations.

For others, however, "business" is not an objectionable word. One extension tobacco specialist pointed out to me at a workshop for "innovative tobacco growers" that those farmers in attendance now treated their operations "more like a business," with the implication that this is positive progress. Roger Quarles explained to me that tobacco has always been, for him, about "economic well-being" and that if something more profitable came along he would get out of tobacco. He then went on to say:

Roger: I think that today's tobacco farmers, the ones that's left in this industry are probably gonna take exactly that approach. And it's gonna be an absolute business to them.
Author: And is that different from, tobacco farmers of the past?
Roger: It probably is a bit. You know we had that huge group of people there that, that would almost tell you they do it because they love it? And I'm not gonna argue whether they did or not.

Quarles, too, sees a changed context for today's farmers, but he accepts this change and acknowledges and then dismisses the nostalgia that others might feel.

Despite different judgments about tobacco farming as "business," there are agreed-upon perceptions of what separates it as a business from earlier contexts of raising it, including the dramatically increased amount of tobacco grown by individual farmers and the resulting changed relationship with the crop. Clarence Gallagher described the difference between large and small tobacco farmers:

> And I saw that when I worked at the warehouse, when I saw somebody that came in [with] 150,000 pounds of tobacco, I mean it looked like it'd been pitched on there with a pitch fork. The bales were just every which a way you know what I mean? Just half tied and, weighed 150 pounds. And a man would come in, and [say], "Well, I got 5,000" and it would almost just be like a—out of a book. I mean his tobacco would be, just every leaf was just in place and just, everything was nice and neat.

Clarence was raising just over 50,000 pounds at the time, so while he was a large grower based on the standards of the past, he was between the two groups that he is describing here. Of course, size really matters because it determines how much attention you can pay to the crop and how it looks; size determines a farmer's ability to perform *pride.* The difference comes back to the aesthetic expression of pride resulting from a physical relationship with the crop and the work.

Frequently, I was told that tobacco production is now about quantity rather than quality—but this was being said at least as far back as 1947. The May issue of the KDA newsletter that year included a discussion of farm practices that had led to increased yields of burley and a comment that strong prices had encouraged a concentration on volume over quality.[35] Martin told me in 2008, "And you know years ago, it was all about quality, and now then it's quantity." This is another indication that if there is a Golden Age period in tobacco production, it has long been a moving one, with each generation looking back at an earlier period as a better one.

Younger, larger farmers—those who might best be described as, and describe themselves as, businessmen—would probably not characterize themselves as "nostalgic." Yet they frequently commented on the ways that tobacco production just wasn't done like it used to be and noted, therefore, that the tobacco was just better in past days. In addition to lamenting a lost relationship with the crop, men's nostalgia is for male-dominated contexts in which men often competed with one another to be the best tobacco men. Instead of cutting tobacco a row over from a buddy or days spent in the male-controlled environment of the tobacco warehouse, many of these farmers spend their days driving alone in their pickups and tractors, overseeing the work of other men—men they perceive as different from themselves based on language, culture, and connections to the crop and the land. While tobacco farmers may still spend much of their time in predominately male environments, a great deal of the time they are interacting not with peers, but with men they view as outsiders and over whom they exercise power, and whom they compensate with cash rather than swapping work.

Not only was there more pride in the work when their fathers did it, according to tobacco farmers, but there was also more respect for the occupation. The changed relationship with the crop and the changed meanings of tobacco are intricately linked. The crop itself, each leaf, was once respected. I asked this of an extension professional who had suggested that larger growers may not follow accepted practices: "And what are those things that don't get done right? Or that the smaller grower does more of, better of?" He responded that when he was growing up,

> every leaf was picked up. Every leaf was valuable. Every leaf was cared for. You didn't step on tobacco on the wagonload. Every plant was, carefully handled. Now, tobacco's walked on as if it were grass on a field. Leaves that fall off of the plants and being handed onto the wagonload, are kicked off at the

> barn, kicked off in the field. Not preserved, not cared for. Tobacco is drug, slung, handled any way to get the job done. Like it was a bag of rock or something, cement. So it's just the care, and the speed now to get the job done. "We gotta get fifty acres in this month." Rather than five acres. So it's just the care and tediousness of, handling the perishable, delicate product.

Tobacco men didn't leave their cut tobacco in the field to get muddy or sunburnt; they didn't step on it or throw it in the baling box. Ultimately this was of course for economic reasons, as it was the individual leaves that paid for the farm, bought the children's shoes, paid for Christmas, and so on.[36] But respect for the crop was intricately tied to respect for the man who grew the crop. Now that the crop is associated with disease and death, and "tobacco farmer" is understood by some as a stigmatized category, the tobacco farmer has, according to some, stopped treating the crop with respect. That's why change in quality matters: the disassembly of the confluence of pride in tobacco and respectability as a man in the community. Jerry Bond described this to me during an on-stage interview at the 2005 Kentucky Folklife Festival:

> There was a time, there was a time when tobacco farmers were proud. And they were proud of their product, and they were proud of their work, and they were—a lot of little towns were built on tobacco farmers—money that tobacco made. But that pride doesn't exist anymore. Tobacco farmers really have become second-class citizens. A lot of people—even today, I noticed down there today [at the festival tobacco farming tent]. People come through and one woman said—I probably shouldn't say this on [stage] —but she said [*nasally voice*] "Well I guess they're gonna show us how to smoke it too" you know. And I thought, "No, ma'am, that's not what we're about. We're only talking about the tradition of tobacco, and the impact it's had on Kentucky. I don't smoke and I don't encourage you to smoke. You know if you want to that's your choice." But you know, she was all up—she was all up in the air, over the fact that there was tobacco being displayed here at this festival. But we—it's just not a prideful thing anymore. You're almost like a drug dealer [*slight laugh*].

According to Jerry, "it's just not a prideful thing anymore": farmers can no longer be proud of a crop that has become stigmatized. The comparison between Berry's description—"In those days, to be recognized as a 'tobacco man' was to be accorded an honor such as other cultures bestowed on the finest hunters or warriors or poets" —and Bond's—"You're almost like a drug dealer"—speaks volumes about the shift in status that tobacco farmers have experienced.

Wendell Berry's essay in *Tobacco Harvest: An Elegy* is representative of a particular thread of tobacco nostalgia, but a nostalgia that is differently expressed depending on the position of the speaker. It is a nostalgia for the way tobacco was once raised, for the men who raised it, and for the respect with which such men were once regarded. Those who once raised tobacco but have now moved on to other jobs or have retired often express this nostalgia through direct judgments of their impressions of today's practices. Current tobacco farmers may also make such judgments, but because they are ever aware of the economic necessity of changing practices they cannot express them directly. Instead, they couch such judgments about today's practices in the perspectives of their fathers, the tobacco men who taught them what they know and who lived in another time.

According to Ray Cashman, the men in his study experienced "a sense of loss coupled with a perceived acceleration of change over the past century that is considered unprecedented and destabilizing."[37] While tobacco farmers commonly express something similar, they also frequently express something close to a sense of marvel, such as Martin's "This technology's great!" Tobacco farmers consistently talk about how things have changed, but they certainly do not want things to be the way they were, at least in terms of the work involved in raising a crop in the current economic structure. Rather, as Stuart Tannock comments: "The 'positively evaluated' past is approached as a source for something now perceived to be missing."[38] Today's tobacco men don't want to return to the difficult work of the past, but they do mourn the respect that was once accorded their fathers. In an ongoing process of loss and gain, they express something like a tradeoff: the loss of the symbolic values of the past in exchange for hoped-for economic value in the present. They embrace the changes and new technologies because of the changed economic contexts in which they raise tobacco, even as they feel less like the tobacco men that their fathers were.

Chapter 7

"Why can't they just grow something else?"

The Challenges of "Replacing" Burley Tobacco

According to the author of a review of James Baker Hall and Wendell Berry's *Tobacco Harvest: An Elegy* in the *Lexington Herald-Leader,* subtitled "Harvesting Our Heritage," "Tobacco growing is going, done in first by health concerns, then by the global economy and cheap imported leaf."[1] The federal tobacco buyout was seen by many as the final blow—even a welcome blow—to what had once been Kentucky's largest cash crop. Headlines proclaimed, "Farmers at the End of Tobacco Road"[2] and "Kentucky Turns the Page on Tobacco."[3] Other headlines suggested that tobacco had been replaced: "Burley Is Just a Memory Now: Bourbon [County] Farmers Turn Their Full Attention to Crops for Farmers Market."[4] According to one news article, "Wine is making a comeback now thanks, in part, to another watershed event, the rapid disappearance of tobacco from the Kentucky economy."[5]

Writing about "narratives of progress and preservation" in Appalachia, Mary Hufford notes that "coal camps in their postindustrial incarnation don't fit into the state's grand narrative."[6] Similarly, tobacco—as a stigmatized tradition—has lost its place in the dominant narrative about Kentucky agriculture. In this chapter, I examine conflicting understandings of "transition" and "diversification" following the 1998 Master Settlement Agreement and the 2004 tobacco buyout. The dominant perspective argues that tobacco production is a thing of the past and that the "transition" period currently taking place in Kentucky agriculture is one in which tobacco farmers have replaced, will replace, or should be replacing tobacco production in favor of "diversified" agriculture: vegetable production, wineries, agritourism, goats, and other alternatives. However, tobacco continues to be profitable for many farmers, and for them, "transition" has meant adjusting to direct contracting with tobacco companies. There are both tangible and intangible obstacles to farm diversification, and "replacing" tobacco is a much more complex issue than the dominant perspective acknowledges.[7] The conflicting meanings of transition and diversification also raise important questions about the uses and meanings of the term *tradition,* a polar term that is often used in contrast with concepts such as innovation and change. Rather, as tobacco production demonstrates, traditional farmers often embrace change, and innovative farmers often expand on existing tradition.

Conflicting Meanings and Uses of "Transition"

Because tobacco symbolism has so dramatically changed—from self-evident center of the agriculture economy to icon to stigmatized crop—the state of Kentucky has a demonstrated interest in changing the image of Kentucky agriculture. Kentucky agriculture as promoted by the Kentucky Department of Agriculture is a "diversified" farm economy ruled by agritourism, vegetable production, and niche products (from salsa to wine and cheese). This is the dominant perspective on the current transition in Kentucky agriculture, a transition in which tobacco is part of the past but not the present. In recent years, the public rhetoric of the KDA has been focused on Kentucky Proud, its successful program to promote Kentucky-produced products. This campaign fits nicely with the contemporary movement that encourages the consumption of locally produced foods as a means of addressing environmental and health concerns and supporting family farms. Bumper-sticker slogans such as "Farmers feed the world" and "Know your food. Hug your local farmer" are representative of this movement, which follows on efforts to save family farms that began during the 1980s farm crisis (most notably through Farm Aid, the organization begun by Willie Nelson and other celebrities).[8] Of course, it is difficult to see how the farmer who continues to raise tobacco—and who may be struggling with financial insecurity and difficulties holding on to his or her land—like the farmer who raises food—fits into this discourse.[9] In order for Kentucky to insert itself into this larger movement to promote the valuing of family farming, the image of tobacco as central to Kentucky agriculture must be replaced with images of healthy food crops.

States made a range of decisions about the allocation of funds resulting from the 1998 Master Settlement Agreement. Kentucky chose to put 50 percent of these funds into agricultural development, and the Governor's Office of Agricultural Policy (GOAP) became the administrative body of the Kentucky Agriculture Development Board, the entity that oversees the dispersal of the funds. The following text formerly appeared on the GOAP website, on a page entitled "Planning for the Future," with a banner that featured a photograph of thoroughbred horses:

> Changes are taking place for tobacco producers because of the changing practices of the big cigarette companies, social and economic pressures from the markets, regulatory agencies and the health industry. Kentucky farm families, political leadership, agricultural organizations, and many others confront the question of *how to best make the adjustment away from* tobacco production in a way that allows farmers to *capture the value of their assets, while adjusting to a sustainable, alternative asset base.* Many producers are leading the way by having already made a successful effort to *replace lost tobacco income.* How can Kentucky help farmers build on the models of our agricultural leaders, maximize the value of their assets, and *explore new opportunities* in production and marketing of agricultural products?[10]

I quote from this text at length because it clearly demonstrates the discourse of transition that has become dominant: that tobacco should and will be the crop of the past and that Kentucky farmers must now "adjust" by "replacing" this traditional crop with "sustainable, alternative" "new opportunities." The argument that many farmers have successfully "replac[ed] lost tobacco income" assumes that there are no symbolic losses to consider alongside economic losses.

The support of projects related to tobacco farming through Agricultural Development funds has been a source of controversy since the fund was formed. For instance, in 2001, the Kentucky Agricultural Development Board rejected a proposal to provide funding to assist tobacco warehouses in their financial struggles when direct contracting began. According to Governor Paul Patton, "I believe that we've got a mandate to use this money to either bring about additional agricultural activity or at least improve existing programs. I don't find this proposal meeting either criteria [*sic*]."[11] The "investment philosophy" of the Agricultural Development Board is as follows: "The Board will invest these funds in innovative proposals that increase net farm income and effect [*sic*] tobacco farmers, tobacco-impacted communities and agriculture across the state by stimulating markets for Kentucky agricultural products, finding new ways to add value to Kentucky agricultural products, and exploring new opportunities for Kentucky farms."[12] This has come to mean that the board will not fund projects that are directly linked to tobacco production, such as building new tobacco barns, but it will fund projects that support the efforts of tobacco farmers in other parts of their operations, such as livestock (buying bulls, building fences, etc.) and alternative agriculture projects. The Agricultural Development Board funds are divided into three types: (1) grant funds that are distributed at the county level through model projects developed by councils set up for this purpose in each county; (2) grant funds distributed for nonmodel projects through a statewide application process; and (3) loans distributed through the Kentucky Agricultural Finance Corporation (KAFC). KAFC funds, unlike the two categories of grant funds, have been used for loans to build tobacco barns through the Agricultural Infrastructure Loan Program, and this is one source of the controversy.

A 2005 editorial in the *Lexington Herald-Leader* began: "Kentucky officials should put an old anti-drug slogan to good use and 'just say no' to tobacco growers."[13] This commentary employs the language of anti-tobacco rhetoric as it concludes, "Overcoming the economic addiction to tobacco is an essential step in overcoming the physical addiction and medical horrors that tobacco wreaks on Kentucky. . . . Putting tobacco settlement money into tobacco-growing would be wrong." A similar argument was put forward in the same paper a year later: "Few have begrudged agriculture the lion's share of the settlement, because Kentucky can never overcome its physical addiction as long it's economically addicted to tobacco. That rationale falls apart when settlement money goes to tobacco."[14] These editorials suggest that the dominant perspective on transition and diversification is based not only on perceptions of how farmers can best assure themselves of a profitable future but on the changed symbolism of tobacco and tobacco farmers. Tobacco production is no longer an embraced segment of the Kentucky agricultural economy; it is now an addiction that must be overcome.

The headlines with which I began this chapter represent the dominant perspective in that they assume that farmers have transitioned away from tobacco production. Tobacco is widely understood as a symbol of Kentucky heritage, a thing of the past. The idea that there are very few tobacco growers left is so strong that even some farmers and agriculture extension agents express surprise at the number of farmers who are still raising tobacco; others acknowledge the invisibility of tobacco farmers today. One agent began keeping a list of farmers who were raising tobacco in 2007 after I asked him to introduce me to growers in his county. He described what he learned from making this list during an interview with me several months later: "Yeah there's way more than, I had guessed. After the buyout and stuff you know a lot of people got out. And so I was guessing that there'd just be, you know, a couple dozen or something." He went on to explain that "just out of curiosity I started [listing] everybody that, you know mentioned they were raising tobacco or I'd drive by and see that they had it or I did a soil test for them or whatever. I started writing their name down and, and I'm still every day, not every day, every week maybe, every couple weeks, hear of somebody new that I didn't realize is raising tobacco that is. And it's, it's easily several dozen still so I'm, yeah I'm surprised, how many there are." Agriculture extension agent Dan Grigson told me, "I think we have more people growing tobacco now than I thought we would, I really do. There are folks who are growing tobacco now that I didn't dream would ever continue to grow tobacco."

In January 2008, the dean of the University of Kentucky College of Agriculture welcomed attendees to production seminars held at a tobacco expo sponsored by the Burley Tobacco Growers Co-operative Association. In his remarks, Dean M. Scott Smith directly addressed the erasure of tobacco from the discursive landscape, commenting that he gets upset when he hears people talk about the "decline" of agriculture generally and of tobacco specifically. He remarked that he particularly resented the many people that he hears say, "Now that we don't grow tobacco." He told those gathered that "we need to educate people that we're here," and he assured listeners that he was "glad to show that the University of Kentucky College of Agriculture is still dedicated to tobacco. We're here as long as you're here." Roger Quarles told me, "Yeah we've heard comments, 'Golly I didn't realize anybody raised tobacco anymore.'" He went on, "I'm not sure how they came to that conclusion. I assume that maybe when they heard about the buyout that that meant—there was probably a misconception that that meant you weren't allowed to grow it anymore and that certainly was never the case." In response to my question about whether he has encountered the perception that "tobacco is gone," agricultural economist Will Snell commented, "Yeah I mean even in this state here we talk about the 'post-tobacco era.' And granted you've lost a lot of participants as part of the program—or part of the industry. But, as I said those, five [or] six thousand farmers that are still left growing, most of them are doing fairly well." The fact that according to the 2007 Census of Agriculture (which had not yet been released at the time of our interview) tobacco was then grown on over eighty-one hundred Kentucky farms only serves to further prove Snell's point: Kentucky is not yet in a "post-tobacco era."

Although images of tobacco production have been replaced with images of diversified farming, actual tobacco production has not been replaced. In 2006, vegetable production represented $20,250,000 in farm receipts, compared with $319,655,000 of tobacco income.[15] According to the 2007 Census of Agriculture, tobacco was grown on 8,113 farms; vegetables were harvested for sale on 2,123 farms; and 428 farms claimed income from agritourism and recreational services. Also in 2007, approximately 2,000 farmers participated in community farmers markets across the state, and over 350 producers sold their produce at auctions.[16] The KDA announced at the end of 2008 that Kentucky Proud was now "1,300-Plus Strong."[17] Also, according to the KDA, in 2008 there were 113 grape producers and 46 wineries in the state.[18] Such statistics indicate a growing number of alternative producers but also reveal that in 2007 there were far more tobacco growers than vegetable or grape producers or farms devoted to agritourism and recreation. Although these numbers have surely continued to grow, tobacco has not yet been replaced because tobacco continues to be profitable for some farmers, and those who want to replace it face both tangible and intangible obstacles. The discourses of transition, diversification, and heritage demonstrate the rhetorical separation of the economic and the symbolic values of tobacco. The heritage rhetoric previously described values the symbolic at the expense of the economic, pushing tobacco into the past and erasing the present economic importance of the crop to many farmers. The dominant discourses of transition and diversification also ignore continued economic value. Conversely, agriculture professionals argue that tobacco is economically irreplaceable without acknowledging its symbolic value.

Will Snell told me, "Well. You know, the classic argument is 'just raise something else.' And you know that's easy to say but, it all comes down to economics and then just, other crops are not as profitable." Replacement does come down to economics, but not to economics alone; there are intangible equivalences that must be met as well. The economic and the symbolic are intricately entwined; when their "reciprocal dependence" is overlooked, "the meaning that proceeds from the totality of which they are part [is] lost."[19] In order to understand the current situation for Kentucky tobacco farmers and to ensure that their voices are heard in decisions about the future, the "reciprocal dependence" of economic and symbolic challenges to replacing tobacco must be untangled.

As Snell points out, tobacco continues to be profitable for many farmers. Such farmers express a perspective on the future of tobacco that conflicts with the dominant perspective described above, as exemplified in the following exchange from an interview I conducted with Clarence Gallagher, who raised twenty-two acres of tobacco in 2007:

> Gallagher: Even though they've had, you've seen all that negative in the paper and everything like this, *tobacco has went up in price* since the [year after the] buyout.
>
> Author: So it's not going anywhere?
>
> Gallagher: Not . . . like I said, I don't think you'll probably ever see tobacco go completely out unless they put a ban on—[unless] you just aren't allowed to smoke in America.

Here, as Gallagher expresses his confidence in the sustainability of tobacco as a cash crop, he also demonstrates his recognition that media reports do not adequately reflect his experiences and observations. Many of those who choose to continue to raise tobacco do so based on a belief that there will always be a market for Kentucky burley—or at least that they will raise it until the market is entirely gone. According to Marlon Waits, who along with his brother and other male family members raised over forty acres of tobacco in 2007, "We're gonna try to stay in it as long as we can make a living at it."

There are a number of factors that support opinions that Kentucky farmers will remain important to the tobacco industry. Because of its flavor and burning qualities, burley is a key ingredient in top-quality American-blend cigarettes. As Dean Wallace of the Council for Burley Tobacco put it during our interview, "I think Philip Morris's Marlboro model, is an unbelievable brand, it's one of the most valuable brands worldwide that's ever been created. And so I think they're gonna be very reluctant to make changes in the mix of that tobacco." The climate of Kentucky is believed, by farmers, agricultural professionals, and seemingly tobacco companies, to be uniquely ideal for curing burley tobacco. While burley can be grown elsewhere, I am told that it cannot be cured—the key to burley quality—anywhere else like it can in Central Kentucky. Robert Pearce, tobacco specialist with the University of Kentucky College of Agriculture, expressed doubts about there being much room for growth in the Kentucky burley market. However, he noted that Kentucky burley would continue to be a "status symbol":

> The "status" meaning that you know, the perception is that this is the best burley tobacco in the world, and so, the status comes from being able to afford the best. And why is ours the best in the world? Well, you know, this is where burley tobacco came from. We set the bar I mean, that's just the way that it—this is where it came from and everybody else has been trying to duplicate what's here. So there is a certain amount of truth to the fact that, you know, it can't be duplicated elsewhere in the world.

"Status symbol" suggests that Kentucky burley has both economic and symbolic value. Pearce cautioned, however: "Now, over the last two decades or so, the amount of imported tobacco going into cigarettes here is increasing. As our prices went up, I mean manufacturers have figured out ways around the inferior quality of some tobaccos and can still produce a product that's acceptable to their end consumer. So, we're not irreplaceable. They may not be able to duplicate what we've got but we're not irreplaceable."

For farmers who have chosen to continue to raise tobacco, *transition* has a very different meaning than the one offered in the dominant rhetoric of transition, in which tobacco is a thing of the past and must be replaced. For these farmers, *transition* refers to the end of the federal tobacco program and the negotiation of new relationships with tobacco companies through direct contracting. According to retired Garrard County extension agent Mike Carter, "You know, 'transition' for many many people, as far as the tobacco buyout and 'what do we do after that' and so on and so forth, is—they've chosen to grow tobacco on a more free-enterprise basis, without the regulations and the price supports and all those kinds of things. So they've transitioned from growing tobacco under a federal program into growing tobacco on the open market." Dan Grigson described a similar meaning for *transition* and added, "But for a lot of folks, they went through a fearful time, you know, 'What am I gonna do?' 'Are they gonna continue to pay me this, or does this look good and they're gonna pay me for two or three years and that's the end of it.' So, a lot of older folks especially, got very nervous, very scared about [what] their future was. As long as they had that quota they knew they had a little something to bring in some income. So that's been the transition for those folks."

An information sheet produced jointly by the colleges of agriculture of the University of Kentucky, North Carolina State University, and the University of Tennessee describes the buyout as "one of the most dramatic changes in any U.S. agricultural policy over the last half century, as tobacco now has the distinction of being the only government-supported commodity to move abruptly to an entirely free-market policy."[20] Rather than meaning a "transition" to other crops, for farmers who choose to continue tobacco production, "transition" means a changing context for raising a crop that they know well, a transition to an environment of uncertainty in which there are no "production controls and no safety nets" for the first time in over seventy years.[21] It also appears to be an ongoing transition period in which growers, without the federal government as intermediary, are becoming increasingly powerless. The use of the term *free market*, though technically accurate, is somewhat misleading since farmers have limited options for the sale of their tobacco. Each year since the buyout farmers have faced increasing scrutiny from tobacco companies; more and more farmers have had their contracts cut (in part or completely); and those who do receive contracts have little control over their ability to fulfill those contracts since the companies determine the pricing structure for each grade and then determine whether or not tobacco meets the criteria.

Diverse has alternate meanings for tobacco farmers as well. Tobacco farmers have in fact always been "diverse," as they are usually also cattle farmers and farmers of hay, feed corn, forages, and other crops. The majority of farmers who have stopped raising tobacco have increased these other traditional farm activities rather than moving to new farming enterprises. Martin Henson learned from his father "not to put all his eggs in one basket," and this proverbial idea seems to be central to the farming philosophies of Kentucky farmers generally. With the exception of the far western portion of the state, Kentucky's fertile but hilly geography has made diversity a necessity. "You have the rolling land, it's not suited for row crops, you're not gonna grow many corn and soybeans on a lot of the land that we have, especially in Central Kentucky and on [into] East Kentucky," Dan Grigson told me. Kentucky farms are much smaller than farms in other parts of the country, and Kentucky farmers have traditionally utilized hillsides for livestock and maximized their flat land for their tobacco, feed corn, and forages. In response to my questions about recent efforts to promote diversification, Roger Quarles assured me that farmers were always involved in multiple activities on the farm: "I guess [those who promote diversification] assumed that people were just sitting around on buckets not doing anything, the time they weren't working in tobacco. For some reason, and I don't know of any farmer that fit that mold. Everyone I ever knew was already doing other things other than tobacco when they had time to do that. They weren't sitting around on the creek bank fishing." Dominant rhetorics of diversification do not recognize this traditional form of diverse farming, however. Instead, as "diverse" shifted discursively to an active process that implies movement away from tobacco—*diversification*—new layers of meaning have been added. I asked a Bourbon County farmer and farm policy activist what she meant when she used the term *diversification,* as she did frequently throughout our interview, and she said: "Well it's more than just having cattle, or just having cattle and tobacco, or just having cattle, tobacco and hay. I mean all those things—you're diversified in having different activities going on on your farm. I think some of the newer ones were—well like aquaculture and—but stepping out. When you diversify, I think more or less you're creating markets, you're finding new ways to farm." She acknowledges the multiple definitions of *diverse* and chooses as her definition the version that echoes the dominant rhetoric of diversification: "finding new ways to farm."[22]

The valuing of innovation through diverse farming is of course not new. In the first decade of the twentieth century, Clarence Poe, editor of *Progressive Farmer* magazine, advocated for diverse farms. According to Lu Ann Jones, "Men like Poe envisioned a modern South unfettered by tradition and open to change. Diversified agriculture would be an important part of its economic base."[23] As John Fraser Hart and Ennis L. Chestang noted in their 1996 article "Turmoil in Tobaccoland," tobacco farmers realized the value of diversified agriculture as early as 1960, as tobacco came under "increasing economic and political attack."[24] At this time, however, diversification in the flue-cured region of North Carolina studied by Hart and Chestang meant large-scale soybean production, vertically integrated poultry operations, contract hog production, and a renaissance of cotton production. Many Kentucky farmers have tried similar alternatives (with the exception of cotton), with varying degrees of success. According to Henry County agriculture agent Steve Moore, efforts to diversify the agriculture economy have repeatedly increased during periods in which tobacco seemed threatened, only to drop off when the tobacco market seemed more stable. He told me that there had been a strawberry program in Henry County as far back as 1959 and went on:

> My [predecessor] I think struggled with a few little things—there were a few people raising cucumbers. And then that stopped, because tobacco came back. And then in the '80s we got into cucumbers again, then we got into bell peppers, we had field days, we had alternative enterprise sessions in the late '80s, where we had a series of meetings where we explored, one at a time, the alternatives. We explored sod farming, beekeeping, you know, goats, sheep, we explored a lot of things in terms of just education, because somebody was trying it around here. But none of those things caught on. Ever. And we did it again in the '90s. And now we're doing it again in the 2000s.

Quarles told me, "I'd say probably there were larger expectations of diversity, by people that didn't realize how difficult the situation would be to go those routes, and I think when you look at the limited success, of all the things that's been tried I think that probably bears that out." Quarles alluded to the difficulties involved in diversifying a farm operation. The obstacles to diversification were frequently relayed to me in the form of narratives about the failures of new crops, particularly vegetable crops grown on contract.

Often, when I asked an extension agent or a farmer if they knew anyone who had tried something "new," I was told a story about someone who had tried bell peppers, Jerusalem artichokes, watermelons, even exotic animals such as ostriches—only to see the experiment end in failure. Former tobacco farmer Phil Sharp responded to my question about whether farmers in his county were trying to replace tobacco with this humorous narrative about a farmer who tried to raise cucumbers: The farmer "said he was pickin' cucumbers. [He] said there was a little one down there the size—right size to pick, and he raised up and lit a cigarette, reached back down there to get it and it done got too big." Like other narrators, Sharp points out here that vegetable crops involve tremendously different marketing circumstances. In the case of cucumbers, peppers, and other vegetables, buyers want a very specific product, and there is only a short window in which to provide it.

Karen Armstrong-Cummins, former managing director of the Commodity Growers Cooperative, said in an oral history interview in 2000, "Everybody's got their green pepper story. 'I tried those green peppers and, you know I ended up with a truckload of *mush*.'" Peppers are in fact the quintessential failed crop in these narratives because so many farmers raised them for brief periods in the 1980s. According to former tobacco farmer Judy Miller, "We raised peppers for a couple of years under contract, and the problem we had with that is whatever you had, if you had green ones they wanted red ones, if you had red ones they wanted green ones." Clarence Gallagher told me a similar story about his experience raising peppers for three or four years in order to supplement his tobacco and cattle income. He told me, "I mean you had a contract with them. But, they would tell you when they needed them. And when we first started raising them well they needed them." He told me they would send postcards that would say, "'These *two* weeks, we'll be taking' like green peppers. And then it wouldn't be long you'd get a card that, 'We'll be wanting the red peppers.' And then 'We want a chocolate pepper.'" According to Clarence, in the beginning, "You could take a pickup load, or two pickup loads or half a pickup load, pretty much get rid of them." Eventually, however, the cards stopped coming, and he wondered, "'What's going on?' Here I've got peppers out here you know needs to be picked." Finally he decided, "'That's it with me.' You know I mean 'I'm not gonna raise these things and then take a harrow out here and harrow them up.' And that's the reason I quit raising peppers."[25]

Dan Grigson told me about a farmer who lost everything: "I won't mention a name but I know a producer in [another] county that invested a tremendous amount of money into horticulture production. He doesn't have any money left today. Because he had one bad year. He had a situation where he had disease and he had dry weather, he had nothing to sell. Almost nothing to sell. So, he's no longer in the farming business today. So it is a huge huge gamble. These traditional enterprises that we have again, you pretty well know you're gonna be pretty safe."

In contrast to dominant rhetorics of diversification, these narratives—representative of others I was told—problematize alternative farming pursuits; in Grigson's story tradition is explicitly valued. These stories speak back to the widely held assumption that tobacco farmers, as a class, should be and are seeking alternatives. They stress the impossibility of replacing tobacco, at both the economic and the symbolic levels, and they critique dominant assumptions that if tobacco farmers simply planted peppers or cucumbers, their problems would be solved. They might also be read as warnings to farmers considering "alternative" crops. As contracting began around 2000, these failed diversification narratives were being told with direct comparisons between experiences with vegetables and fears about raising tobacco on contract. Former Burley Co-op CEO Danny McKinney pointed this out explicitly in an oral history interview in 2000, as direct contracting of tobacco was just beginning: "Our experience with contracting has been that the first year or two it will work pretty good. When the green pepper market comes to town, they'll sign the farmers up—the first year or so it works pretty good, but after that, whatever you've got they don't want and what they want, you don't have. And that's the problem we have with contracting." In contrast to these narratives, the dominant rhetoric of diversification values "newness" and "creativity" over tradition. To be "diverse" a farmer must move beyond traditional ways of farming, must be progressive and innovative—even risky. As value is placed on the innovative farmer willing to take risks, the traditional farmer is devalued. However, the seeming dichotomy of tradition and innovation demands further attention.

Complicating "Tradition"

According to Wendell Berry, "For some people, some Kentucky newspaper editors among them, the new infamy of tobacco [after the surgeon general's report] legitimized the old prejudice against farmers and country people. The growers were condemned along with the crop, even though the farms that produced tobacco had always been diversified and were contributing significantly to the food supply."[26] Here, Berry connects the conflicting meanings of diversification with the newly stigmatized category of "tobacco farmer" and the more generalized stigma that farmers and rural people face as their numbers have dwindled in an increasingly urban/suburban America. He suggests stereotypes of farmers as stubborn and old-fashioned. Farmers who continue to raise tobacco are often portrayed as particularly resistant to change. For instance, a November 2007 report on the PBS news program *NOW* featured an effort by one Virginia organization, Appalachian Sustainable Development, to "persuade farmers to change from growing tobacco to growing organic fruits and vegetables."[27] This program stressed the difficulties involved in "confront[ing] centuries of tobacco tradition" and "convincing tobacco farmers to break with what used to be tried and true," echoing the dominant rhetoric of diversification by putting tobacco farmers in a position of direct opposition to the organic foods movement.

Yet tobacco farmers have historically been quite willing to adopt innovations. The many technological innovations that burley farmers rapidly adopted over the years include maleic hydrazyde in the 1960s, baling tobacco in the early 1980s, and float beds in the 1990s. Farmers moved from tying tobacco into hands to compressing it into bales rapidly; once the technology was developed, it effectively took just three years to implement. The following story, told to me by former warehouseman Ben Crain, is representative of many stories I heard about fathers and grandfathers and their willingness to change. He was telling me about his grandfather, who passed away in the early 1960s, and I asked if he was around when sucker control was introduced.

> Yes he was and it's very interesting that you should bring that up. I guess I was the only grandson at the time, or the oldest grandson. And that was in the '50s now I'm talking about. So I was always given a half-acre of tobacco that was my half-acre that I shared with my grandfather. So I'd get my half interest of a half-acre. And in roughly the mid-'50s, the advent of the sucker control came about. Dad planted the seed, "Why don't we try it on Ben's crop?" So, that was finally agreed upon. My grandfather said we were gonna ruin it. We were gonna ruin that crop of tobacco. And it was a good one.

Crain went on to describe how, after they had sprayed it with sucker control (or maleic hydrazyde), "the tobacco on either side of my half-acre was just full of suckers," while "my half-acre did not have a sucker on it and it was absolutely beautiful and yellow when we cut it. And I think again we saw the handwriting on the wall there. No more suckers." I asked if that meant his grandfather was in agreement with the use of sucker control after that, and he said, "Oh yeah, after that. Yeah absolutely." According to Crain, his grandfather had only to see that he would benefit from an innovation such as sucker control, and he fully embraced it. Why, then, are farmers stereotyped as "traditional" and resistant to innovation?

Martin Henson told me in an interview, "I get asked every now and then 'Why are you raising, still raising tobacco?' 'Tradition' is what I tell them." He also told me, "I'm old-fashioned. My wife says 'You can't teach an old dog new tricks' she says 'He does not like change' and I don't, I don't." "Tradition" is of course a central concept for folklorists, a complex concept that we and other scholars have increasingly thought, as Dan Ben-Amos puts it, not just *with* but *about*.[28] Despite our efforts to rethink our use of this term, *tradition* continues to be employed in both scholarly and everyday usage as what Kenneth Burke calls a polar term, a term with implied opposites such as change, creativity, innovation, dynamism, and modernity.[29] As recently as 2005, James Bau Graves described tradition as the "synergetic opposite" of innovation and "the rock for innovation to push against."[30] Scholars such as Dell Hymes[31] and Richard Handler and Jocelyn Linnekin[32] have complicated our understanding of tradition, treating it as process rather than an object. However, we've not yet asked enough questions about the vernacular uses and meanings of *tradition*. What did Martin mean, for instance, when he told me that "tradition" is what he tells people when asked why he continues to raise tobacco?

It was in this same interview that, as I discussed previously, Martin exclaimed, "This technology's great!" From a perspective of understanding tradition as static, rather than as always in a process of change, this seems to contradict his statement about not wanting to change. However, the contexts of each of Martin's statements not only are crucial to aiding our understanding of what appears to be a contradiction but also suggest the need for further examination of the meanings of the vernacular usage of *tradition*. His exclamation "This technology's great!" occurred following his story about his shift from starting tobacco plants in the ground to the use of polystyrene seed trays floated in water beds. His description of himself as "old-fashioned" and resistant to change came during a discussion of the difference between growing tobacco versus raising vegetable crops and in response to my question, "You think it'd be hard to transition into something else?" While Martin praises technology within the context of tobacco-related innovations, he refers to himself as unwilling to change when discussing the move that some farmers are making to vegetable crops and in response to a direct question about whether he would find it difficult to join them.

According to Henry Glassie, "Change and tradition are commonly coupled, in chat and chapter titles, as antonyms. But tradition is the opposite of only one kind of change: that in which disruption is so complete that the new cannot be read as an innovative adaptation of the old."[33] As I learned from Martin and others, change is understood as part of the tradition of raising tobacco rather than as a force to be fought. Some of the tobacco farmers with whom I interacted were described by others, particularly by extension agents, as particularly innovative. These were farmers who worked closely with the Extension Service, testing new technologies such as mechanical harvesters, nonconventional curing structures, and conservation tillage equipment. According to George Duncan, "Farmers are good innovators on their own," and their ideas often spur university research. Major innovations such as float beds serve as examples. Growers became interested in float beds, and the university responded by beginning research in order to meet their needs. For Martin and other tobacco growers, while many of the changes that have taken place in tobacco have been accepted as part of the tradition, growing tomatoes would be a complete disruption of their farming tradition.

In common usage *tradition* often refers to outmoded and old-fashioned ways and is understood as a restrictive and emotionally driven force. Adherence to tradition is often associated with backward rather than forward thinking and aligned with emotion rather than the realm of the practical or the tangible. While this statement may appear to reinforce a dichotomous understanding of the emotional versus the practical, I point to this dichotomy in order to problematize rather than reinforce it. As Handler and Linnekin have argued, tradition is not "a core of inherited cultural traits whose continuity and boundedness are analogous to that of a natural object" but is instead "a wholly symbolic construction."[34] However, this does not negate vernacular usage of the term to encompass aspects of the intangible as well as the tangible.

When Martin told me he continues to raise tobacco because "it is tradition," he was not telling me that he was emotionally bound to the crop or even, in Handler and Linneken's terms, to his "symbolic construction" of the crop as it represents a way of life. He may be, but tobacco is also both what he knows and what he is materially equipped to grow, and he is bound to the crop by material and economic ties that are part of his understanding of tobacco as "traditional" for him. This became even more evident when I asked for clarification about his use of *tradition.* I said, "You said earlier that people, when people ask why you grow tobacco you say 'tradition'—can you say more about what you mean by that?" Martin responded, "Oh it's just . . . sum it up in one word [*laughing*]. Well, I still use it—[for] income" [*overlapping*]. "But is it just about income?" "But I just tell people 'tradition,' you know, cause, you know. Uh, income. And I still, I've got neighbors that go into vegetables and stuff like that. I still think tobacco's easier than vegetables. I haven't got into the vegetables, but I don't like to pick beans. Be hard on my back [*slight laugh*]." I asked a question that would be hard for anyone to answer—to articulate the meaning of tradition—and therefore it is no surprise that Martin had trouble answering it, as evidenced by his uncharacteristic halted speech. But what he did say—and what I didn't hear until I transcribed it—is instructive. I was surprised and a bit embarrassed to hear my own resistance to Martin's linking of "tradition" with income, when I asked, "But is it just about income?" It is obvious to me now that I wanted Martin to talk about intangible aspects of tradition—perhaps to wax on about his emotional relationship with tobacco—and instead he was telling me that tobacco is important for the income it continues to provide and that its economic value is inseparable from his understanding of it as a "tradition." For him, the tangible and intangible values of the crop are intricately tied together, making economic replacement difficult and symbolic replacement perhaps impossible.[35]

I asked Franklin County extension agent Keenan Bishop about differences between raising tobacco and raising other crops, and he responded:

> Everything [is different], because, with tobacco, you know even if, you didn't like it or, you weren't the best person at it, you know you grew up saturated in it—your daddy did it, his daddy did it, all your uncles did it, so you couldn't help but learn how to do it, how to do it right. So, you know, you grew up in that atmosphere that culture where, you could do it with your eyes closed kind of thing. And all the farms were suited for it, all the farms have tobacco barns and all the farms, you know, have your little patches along the creeks or the ridge tops. You know, to grow the crop it was a no-brainer kind of thing. So to do anything other than tobacco, you've got to not only relearn, but you've got to regear for it. You know, you may be able to use some of the equipment but probably not.

His response encompasses intangible traditional knowledge as well as tangible things—land, equipment—that underlie an understanding of tobacco as a traditional crop.

There are of course farmers who have successfully reinvented their farming operations, and their experiences further illustrate challenges to diversification. One such farmer is Kevan Evans, who once raised tobacco and cattle and who now, in partnership with his daughter, runs an agritourism operation in Scott County. The Evans farm has been intentionally remade as a public space, including an orchard, farm stand, play area, and concession stand. This venture began in the 1990s, when Kevan asked his daughter Jenny, in high school at the time, if she would like to raise vegetables for the newly opened Scott County farmers market in order to earn a little money for college. They began with two acres of vegetables. When I interviewed them in September 2007, they were farming about fifteen acres of vegetables, and their orchards included eight acres of apples, two of pears, and about three of peaches, and they had recently added blueberries, blackberries, and raspberries. I asked Kevan what kinds of knowledge had transferred from his past farming experience, and he replied:

> Well like spraying, watering, you know, and the soil fertility and those type of things and when you're scouting. Because you know, we don't just random put a spray on, we're scouting to look for bugs or look for disease problems and those type things so. This was just a carryover from some of those others—I mean you had to learn new diseases, you know, new bugs and those type of things. A lot of that stuff just carried over from other crops. So it wasn't that bad. But you did have to learn, you know, "When is this apple ripe?" and "What do I need to be doing?" "How do I need to trim that tree to do what I want it to do?" So, I mean we've made a lot of errors that way—a lot of it's trial and error.

Kevan made it clear that although a great deal of previous knowledge was transferable to new crops, there is also a great deal of new knowledge that must be obtained: How do you know when an apple is ripe, when to pick it, how to trim the trees, how to make cider? What about diseases? One extension specialist told me: "It's very challenging with vegetables because, to make good money at it, you have to basically do retail sales. Retail means a real big commitment to, number one growing a wide variety, like whereas you might be growing two varieties of tobacco, you have to grow, you know, five different varieties of thirty different vegetables to make it [in] farmers market sales or restaurant sales." While certainly tobacco farmers can learn all of these things (and many have), they must first embrace a learning process that is very different from the primary means through which they learned to raise tobacco. As Keenan Bishop pointed out, most learned to raise tobacco slowly over a period of many years; they learned by doing and from other farmers, as well as through more formal sources such as the Extension Service. Moving to new crops requires a commitment to an accelerated learning process and an acceptance of the possibility of financial failure.

Like Keenan Bishop, Kevan Evans stressed a significant economic challenge: "We're doing this, it's not generational—if I do something I gotta have a new piece of equipment because it's not there." Because farm equipment is most often inherited, shared within the family, and/or accumulated over years, a changed operation means the need to purchase new equipment, remodel barns and other structures, and otherwise retool the farm—an enormously expensive proposition. Land is of course another resource that affects a farmer's ability to make changes in his or her operation, and as I have already discussed, Kentucky farms are traditionally small, and with the exception of the western portion of the state, most are predominantly hilly land not suited for large acreages of row crops. A small farm imposes limits on diversity because there is less land to farm and less opportunity for crop and pasture rotation. A changed farming operation involves an investment of time as well as money (and time *is* money, as the saying goes): waiting for orchards or grape vines to develop, for land to lay fallow until it can be sown with organic crops, to make mistakes and learn to get it right. Such economic factors represent tangible aspects of tobacco as a farming tradition.

A related challenge is less tangible: a changed operation means upsetting the balance on the farm that has been maintained for decades by the combination of raising tobacco and cattle together. The rhythm of daily life organized around tobacco and cattle has become traditionalized.[36] Tobacco production requires attention year-round, and producers and their families live by the tobacco calendar, knowing each month, each day, what needs to be done with the tobacco. Transferring this knowledge to crops with completely different production cycles and marketing needs is a matter of not only gaining new knowledge but altering patterns and rhythms of everyday life, as it has long been lived. A crop like bell peppers—the quintessential failed crop—involves a wholly different production cycle and therefore new patterns of daily life on the farm.

The failed diversification narratives, as well as a more general consideration of vegetable crops, orchards, and other alternatives, also reveal one of the strongest ways in which tangible and intangible challenges to diversification are intricately entwined: marketing. The tobacco market is relatively stable. According to Roger Quarles, "You know most markets work in cycles, that's one thing about the tobacco market it's very steady, it plods along, and it doesn't have these huge swings."

> Cattle were up high a year ago, now they're not worth raising according to some people. And goats fit the same thing. Any

> crop you pick will almost fit those cycles but, tobacco—that is it's a very, a very dependable crop, to put out I mean. It's one of the very few I know of, that you can have some reasonable expectation about what your income is gonna be this coming year. I mean you're not gonna get enough to, you know, get rich in any one year but, but then you know, you're not gonna go broke either. You're not subject to the market swings that some of these other crops are.

While some would disagree with Quarles's confidence in the stability of tobacco (pointing, for instance, to the drastic quota cuts in the late 1990s, as well as the recent contract cuts), prices have historically gone up and down incrementally as compared with other farm products. Another marketing issue is less debatable: because such crops are perishable, they require immediate access to a market. If that market is not ready to receive the crop, it will be lost—as pointed out in many failed diversification narratives. Although tobacco requires specific knowledge about when it should be set, topped, housed, and stripped, once it is stripped, graded, and baled it can sit in the barn until it is time to sell it. At times, farmers have held on to their unsold tobacco until the following year if necessary.

But perhaps an even greater challenge than getting the crop to market is that the alternatives to tobacco that are stressed in the dominant rhetoric of diversification all require finding, developing, and maintaining a market and producing products that meet the expectations of that market. Tobacco farmers never had to find and develop a market. The market came to them, and they had the traditional knowledge necessary to meet the demands of that market. There is a world of difference between marketing tobacco—and cattle, for that matter—and marketing alternative crops. According to Kevan:

> Because you know with tobacco if you did what you're supposed to do, you had a guaranteed product, a guaranteed sale for what you had. And all this is, is marketing. I mean it's just, being a farmer, a lot of people can raise it, but being able to sell it. [My daughter] Jenny does an excellent job with that and, if we can get a good enough product here, and draw enough people here to buy it, then that's where your tourism, you know entertainment type stuff [is]. A lot of that brings

them here to do that. Then you've got the product there that you can move.

The tobacco program alleviated a farmer's need to find a market for his or her crop, because if a tobacco buyer did not buy the tobacco at auction, it would go into the pool, and the farmer received the minimum price for it. Marlon Waits told me, "As long as it was [a] pretty decent-quality crop you had a support price. If the companies didn't want it you still, you were guaranteed so much." Alternative farmer Susan Harkins described the contrast this way: "And that's something that tobacco did to the farmer. The beauty of the tobacco program is that it allowed the farmer to just focus on raising a product, and they didn't have to think beyond, stripping it down, bundling it up, carrying it up to market. And laying it somewhere on a pallet. They didn't have to do anything else." She went on to say that the tobacco program "was a gift, but it was a curse": "But when you get into these smaller, items, vegetables and shrimp and different things—there is no market—you have to go create your market."

Moving from tobacco to other crops requires learning entirely new skills—nonfarming skills—that enable a farmer to develop or create a market, or it requires having a partner with those skills. In a sense, diversification requires two occupations: *entrepreneur* and *farmer.*[37] Kevan described the need for both a different set of skills and the labor of two people. He handled the difficulty of a new farming venture by being able to focus fully on the agricultural aspects; he was free to do so only because his daughter, Jenny, was handling the marketing. Because Jenny is a member of his family, he was more able to take the financial risk involved in completely restructuring the farm operation. I asked if he'd be doing what he's doing if Jenny did not work with him, and he responded, "Don't know—don't know if I'd do that or not," because "you know it's almost a full-time job raising it, and it's almost a full-time job marketing it. And that would have been tough without Jenny, or somebody else in that position. To do that, to be able to do both. Especially on a retail basis. I mean, I could raise it and sell it wholesale, but it's really—you don't know if you can make enough money doing it wholesale."

I was told many times that there is no single answer to the replacement of tobacco. Instead, each farmer must find or create her or his own niche market. However, as Alice Baesler pointed out, such niches may prove difficult to maintain. "One fellow in Midway that raised tomatoes, grew all the tomatoes you would need for Midway. So, and strawberries—you know 'you pick' strawberries, you can only have so many of those patches. Or then nobody makes a profit, so the biggest thing is coming up with an innovative idea, of 'What I can do that's a little bit different.'" Another category of failed diversification narratives I was told featured the kind of stories that Alice alluded to, in which after a farmer successfully raised a new crop—such as meat goats or pumpkins for picking by local consumers—neighboring farmers attempted to emulate his or her success. In these stories, the result was a failure for all because there is only so much demand in the local economy for such alternatives. A flooded market is a challenge that one rarely hears about in the dominant discourse of diversification or in the push for farmers to raise products for local markets. Local economies must be able to absorb the products that farmers begin to produce if farmers are to be successful.

The more tangible entrepreneurial skills of finding and maintaining a market are absolutely necessary for diversifying farmers. Equally important, however, is the aesthetic system needed in order to consistently provide a product that conforms to consumer expectations. According to Kevan Evans, "Jenny and Sue, my wife, were real sticklers of quality. I couldn't take a second, any vegetable that was a second. You know we didn't put it on the shelf, they just didn't allow it. 'Oh, Dad, this is not worth it.' You know, and we'd take it off. You know I was from that old school where, you know okay, 'It's eatable!' Edible, you know you should be able to put it on the shelf. But the customer has already gotten used to Kroger and stuff you know it just, it had to be a just number-one grade-A apple." Here, Kevan described what he sees as a major break with tradition, made clear when he described himself as "from that old school." The "school" to which he belonged is gendered, yet this is not a simple case of gender stereotypes, of women being more concerned with appearance than men. This is instead about the inseparability of the symbolic and the economic value—intangible knowledge of an aesthetic system as it is applied to a tangible object—of tobacco production. Tobacco-man knowledge is a different system of knowledge than what is needed in order to produce crops that have often, if not always, been grown by women.

Under the tobacco program, farmers were glad to have a guaranteed minimum price, as Harkins points out, but most worked to produce tobacco that would bring the *maximum* price, so while they did not need to focus on finding a market, they had to do more than "just focus on raising a product." They focused on raising the highest-quality tobacco of the grades and types that companies wanted. At every stage of production there are decisions to be made that will affect the tobacco that the farmer will sell—what varieties to choose; what chemicals to use and how much; when to plant the beds and when to set the plants; when to top, cut, house, and strip the tobacco; the weather on the day of sale; and so forth. This is the *tobacco-man* knowledge described throughout this book. The skills needed to market tobacco in the post-buyout environment are largely the same, although as Peter Benson points out, "For farmers, contracts mean marketing access but also new pressures related to keeping tobacco clean. The trip to Philip Morris's contract station" "is fraught with uncertainty."[38]

Tobacco farmers have traditionally taken a great deal of pride in how their tobacco looks—some say in order to improve the price, others say just for the pride of it. Tying a "pretty" hand of tobacco was a source of male pride, and when the move from hands to bales took place, the traditional aesthetics transferred to the new method of packaging. Many farmers continue to strive for an aesthetic ideal when baling their tobacco, just as they did when they tied hands and saw that they were carefully laid onto baskets. Noel Wise said in reference to a bale of tobacco during our interview, "Boy it was a pretty thing to look at, you know he'd put a lot of time into it."

Raising new crops involves more than the transfer of an aesthetic system—it requires learning an entirely new system; the tobacco aesthetic system cannot simply be transferred to vegetables and fruits as it was from hands to bales. Martin Henson and I discussed farmers who were moving to vegetable production, and he told me: "I'd be afraid that I'd raise a bunch of tomatoes or something and they wouldn't—they'd have a speck on them or something [and] they wouldn't take them [*laugh*]. I mean you know. When you go to the market, you're—the producer is, the low man on the totem pole."

According to Martin, "You gotta do what the customer wants." And yet he recognizes that it is no different for him as a tobacco farmer, as he went on to say, "But of course you know the customer's always right. You know, that's the way it is on everything. The companies they're always right, when we take our tobacco there. That's it, bottom line." Martin and other tobacco farmers, as described by Keenan Bishop, have lived and breathed the knowledge of how to raise tobacco so that it looks like it should. As Kevan Evans described (and Martin implied), new endeavors require the attainment and application of new aesthetic systems—intangible systems applied to tangible objects such as tomatoes in order to provide desirable products for a new type of customer.

Jenny Evans's role at Evans Orchard provides an important example of gendered aspects of the current transition period in Kentucky farming. Women are playing important roles on farms involved in experimental diversification—from vegetables and flowers for farmers markets to wineries and agritourism enterprises.[39] Importantly, however, these women also represent the continuation of a wider gendered agricultural history, as their efforts can be understood as an extension of women's traditional roles on the farm. There is a long history of women selling the products of their labor—things they were already producing—when they had a surplus and/or money was needed to cover expenses. As Lorraine Garkovich and Janet Bokemeier note, "Women often used surpluses from goods produced for household consumption to produce goods they could either sell to local store owners directly for cash or trade to them for in-kind purchases."[40] Tobacco farmer and warehouseman Jerry Rankin told me, "My mother sold eggs every weekend" at the grocery store. "She would sell frying chickens. And cream and milk. And have money left over after she bought her groceries."

Women saw their families through bad crop years and times as tough as the Great Depression. As farms industrialized and American consumerism became increasingly centralized, women's productivity on farms declined.[41] Diversified agriculture in Kentucky and elsewhere seems to be an example of a reversal, however small-scaled it may be, as some farmers attempt to move what was once viewed as supplemental to the center of their farm operations. Even such a seemingly nontraditional operation as Evans Orchards can be understood as an extension of tradition, as Jenny and her mother make jams, pies, and other food products for sale. When I interviewed Charlene and Charles Long in 2005, Charlene told me about a couple who began to raise vegetables on a whim only to see their effort take off into a successful new operation:

> We've got one over on [Highway] 31E, they, he started it as a little crop for his wife. She wanted to make a little extra money. Well she raised tomatoes and vegetables. Well . . . it boomed out for them. They built—they've got [a] roadside

> stand and they've built this big thing and they'll have a big Fall Day and everything. They sell tomatoes to all the places around. They sell Olive Garden stuff, they raise pumpkins, and everything, and he said "Oh there was a whole lot more money in that than tobacco and really not much more work."

I asked if this couple was surprised that their switch had worked out so well, and Charlene told me she thought so, because "he gave her an acre out there to work the first year. And then the next year she wanted more, and after he saw what she was making and all, he decided to go into it big time." At this point Charles said, "We are a people that's—we do not want to change. But if you, if you can show us that change is good we'll—we will change." Like the story of the Evans Orchards, in this example what began as a side business for a woman in the family led to something new.

Although it is not always articulated overtly, gender has become mapped onto crops. Vegetables and flowers are aligned with the household and grown in "gardens"—as opposed to "fields" in which "farmers" work—which have often, although certainly not always, been tended by women as they supply food for the family (sustenance) and, in times of surplus, to be sold for (extra) income. The gendering of crops has long been acknowledged in the literature on gender and development and has served as a basis for agricultural policy making in "developing" nations. Often, an understanding of vernacular conceptions of "women's crops" and "men's crops" has been a means used to address women's vulnerabilities in such contexts.[42] Here, I think it can be useful for understanding both the important role that women are playing in the current agricultural economy and the complications of "replacing" tobacco. Tobacco farmers are performing a traditionalized masculinity as they choose to continue to raise a crop that has been gendered male, tobacco, versus moving to crops that symbolize the feminine household sphere. "Diversifying" requires a new performed identity in addition to new knowledge, skills, equipment, and so on. This new performed identity must replace the traditionalized performance of masculinity of fathers and grandfathers, the models of the Golden Age of the tobacco-man masculinity previously described.[43]

The stories of Evans Orchard and of the challenges faced by tobacco farmers suggest that the tradition/innovation binary is problematic, as tobacco farmers—"traditional" farmers—are often described as innovative, even as "diversified" farmers (those categorized as "innovative" based on the dominant discourses) are often building on gendered traditions. *Both* tradition and innovation are central to *both* tobacco and the new diversified farming, yet value is placed on "innovation." It is for this reason that tobacco farming is categorized as traditional and diversified farming as innovative, even though neither fits neatly into one category.

As I have suggested, understanding tradition and innovation as binary opposites is problematic. So, too, is such an understanding of traditional and diversified farmers. In many cases they are quite distinct categories, but there are also many exceptions. Shell Farms in Garrard County is one such example. In addition to about 115 acres of tobacco in 2007, the three generations of Shell men mentioned elsewhere in this book also raised about 300 acres of tobacco plants for sale to other farmers, they usually have about eight hundred head of cattle, and they grow annual flowers and vegetables that they sell both out of their greenhouses and on a wholesale basis to area nurseries. Such examples exemplify yet again how this transition is playing out on Kentucky farms, as this family continues to raise Kentucky's traditional crop, burley tobacco, at the same time that they diversify their operation in innovative ways. The Shell farm is not your grandmother's garden.

Not only is it not your grandmother's garden, but according to dominant rhetorics of transition and diversification, Kentucky's new agricultural landscape has replaced your grandfather's farm. The Evans farm, representative of the model of the dominant discourse of diversification, is one to which children and families can proudly be invited—unlike a tobacco farm. Tobacco has been tarnished through stigma, and the state of Kentucky is clearly interested in presenting a new, cleaner face to the nation.

Conclusion

Burley

Much has changed in the Burley Belt since the major period of my research, the 2007 crop year. Tobacco companies—with Philip Morris in the lead as the largest burley buyer—have stopped offering incentives to farmers who fulfill their contracts, they have cut many growers' contracts significantly or entirely and increased others, and they have begun to penalize farmers who do not package their burley in large bales. Growers and tobacco specialists express the belief that the tobacco companies—again with Philip Morris in the lead—are culling growers whose tobacco quality does not meet their standards. Several of the farmers I came to know have voluntarily stopped raising tobacco—including Martin Henson, who raised his last crop in 2010. When I asked him why he had stopped raising burley tobacco, Martin described his last two crops and told me there was no joy in it anymore. Both had been good crops through the summer and when he cut and housed them, but then he watched as both crops refused to cure in the fall weather. When he brought his tobacco to the receiving station, he was offered prices that in some cases didn't cover what he had put into raising the tobacco. He was reminded of the changed circumstances of selling tobacco post-buyout, absent government tobacco graders and support prices, as the same tobacco brought dramatically different prices on different trips to the receiving station. According to Martin, there is no joy in raising a good crop, watching it dry up, and being subject to the whims of tobacco company buyers. Like many other farmers who've quit raising tobacco in this transition period, Martin now concentrates on his cattle—he's increased his herds and now raises corn and hay where he once raised tobacco.

While this book has focused on particular periods of transition, I hope that I have demonstrated that burley tobacco culture should be viewed as having always been in transition, from the time inhabitants of the Americas first "came across [the plant] about 18,000 years ago," to the beginning of cultivation of the plant sometime between 5000 and 3000 B.C.E,[1] to John Rolfe's first attempt to plant tobacco at Jamestown in 1612, through the establishment of an American farming tradition stretching from colonial times to today. I set out to do ethnographic fieldwork with tobacco farmers only to realize that I needed to examine both contemporary and historical contexts in order to understand the changes that farmers have experienced and continue to experience. This, in turn, provided the structure of this book, the unraveling of multiple threads of change and their meanings to farmers.

One such assemblage of threads includes changes in farm technologies, labor, and tobacco marketing practices. This "thirteen-month crop" requires the farmer's involvement in the crop year-round, as the rhythm of daily life entails involvement in the current and sometimes also the upcoming crop. The mastery of particular skills is crucial at every stage of production: from preparing the ground; to knowing when to *set, top, cut,* and *house* tobacco and when the crop is *in case* and ready to *strip;* to packaging cured tobacco for the market in particular ways that ensure a *pretty* crop that will bring a high price. I often heard the refrain "No two years are the same." This mastery of tobacco-man skills centers not just on the ability to engage in the same practices year after year but on the ability to anticipate and adapt to changing circumstances of weather, pests, diseases, and so on. Such adaptability is understood as part of the tradition of raising tobacco, and it is this mastery and its role in the rhythm of daily life that farmers are referring to when they call tobacco farming a tradition, but they are also referring to having the specialized equipment, structures, and land required in order to carry out these practices. Also part of the tradition are the many technological and marketing changes that farmers have accepted, including maleic hydrazide, or sucker control, and the shifts from hands to bales and from plant beds to float beds. The end of the quota system and the tobacco auction is not so neatly categorized. Some growers have accepted it as yet another change within the tradition and have transitioned to raising tobacco under these new circumstances. For others—both some who continue to raise tobacco and some who do not—the end of the tobacco program is a "disruption . . . so complete that [it] cannot be read as an innovative adaptation of the old."[2] All those who continue to raise burley tobacco are doing so in a different world.

The changed political meanings of tobacco proved to be an equally important thread to unravel. Once I began my fieldwork with farmers, I realized that assumptions I did not recognize I held were being challenged—such is the nature and value of ethnography. For instance, I lived in Kentucky in the late 1990s and early 2000s, and my scant knowledge of what was happening in that period—the Master Settlement Agreement, a potential tobacco buyout—came from my casual awareness of public discourses about the "situation" for tobacco at the time. Looking back, headlines like "Farmers Kicking Tobacco Habit—Success of Alternative Crops Eases Burley's Hold on Region"[3] led me to assume that tobacco farmers were all looking for alternative crops with which to "replace" tobacco. Once I realized that this was not the case, once I met farmer after farmer with no intention of "diversifying," I realized that it was important to investigate the sources of my assumptions and, therefore, to examine the discrepancies between public discourses about Kentucky burley production and what I was learning from farmers with whom I interacted throughout my fieldwork.

Once I noticed the absence of images of tobacco in the offices and on the website of the Kentucky Department of Agriculture, I turned to its newsletter in order to find out when and how tobacco disappeared from the state's representation of the Kentucky agricultural landscape. As the state agency most obviously devoted to agricultural policy and marketing, the KDA plays a unique role in creating the public face of Kentucky agriculture. The newsletter is a primary site for the purposeful articulation to the public of the priorities and perspectives of the agency. As I read through every issue, a clear pattern of changes emerged. In the 1940s, the KDA reported on tobacco in what I have described as a self-evident fashion: it was the largest cash crop; more smokers meant economic growth. By the 1950s and 1960s threats against the industry from research linking smoking with disease were mounting, and the rhetoric of the newsletters shifted to reflect first indirect and then explicit defenses of tobacco. Importantly, the shifting tactics of the tobacco companies as they established the grounds for debate about tobacco can be seen in these pages. The KDA deployed a defense of tobacco based on its economic importance, followed by a linkage of this economic argument with tobacco's economic heritage. The thread of heritage is woven so thoroughly through the story of tobacco that it is difficult to untangle. Heritage discourses are ultimately defensive discourses, deployed under perceptions of threat. According to Barbara Kirshenblatt-Gimblett, heritage "depends on display to give dying economies and dead sites a second life as exhibitions of themselves."[4] In other words, *heritage* is deployed when economic value is understood to be going or gone, and symbolic value is understood to be all that remains. The economic value is understood to lay not in present significance but in the past—*this is who we were*—unless new economic value can be generated through heritage, for instance, in the form of museums, tours, and trails.

In the ensuing decades, tobacco was increasingly threatened by the growing evidence of tobacco-related illnesses and by a growing public awareness that the industry had hidden such evidence from the public for decades, and heritage became a means of defense deployed by the tobacco companies and the state. Eventually, however, heritage became disconnected from the economic arguments as tobacco was increasingly represented as a "way of life" and tobacco work was reproduced as art. Through such representations, tobacco production became something to look back upon; tobacco production in the present became nearly invisible. The heritage discourses of the tobacco industry and the state specifically centered on the tobacco farmer as an American icon in an attempt to put an empathetic face on the industry. Not only did this fail to recover tobacco, but it helped to secure the commensurability between farmers and tobacco companies in public opinion. By the 1990s, images and stories of tobacco appeared less and less in the KDA newsletters. Heritage discourses were largely abandoned by the KDA by the turn of the twenty-first century, and the state instead increasingly promoted an image of a diversified agriculture economy—*this is who we are now.* By the early years of the twenty-first century, major changes in agriculture policy such as the tobacco buyout were not reported on. Tobacco had been replaced in the image Kentucky promoted of itself, even if it had not been replaced on over thirty thousand Kentucky farms at the time of the buyout.

Heritage is also a stigmatizing discourse. Kirshenblatt-Gimblett defines heritage as "the transvaluation of the obsolete, the mistaken, the outmoded, the dead, and the defunct."[5] This means that as the tobacco farmer was cast as the central character in the rhetoric of tobacco heritage, farmers who raise tobacco came to be understood as part of the past as well—obsolete, mistaken, outmoded, dead, defunct. If tobacco is in the past, then those farmers who continue to raise it must be old-fashioned, resistant to change, stuck living in another era. Heritage discourses combined with the new associations of tobacco with death and disease combined to make it easy to ask, *why don't they just grow something else?*

James F. Abrams suggests that not all expressions of heritage are the same, however, and that "the vernacular continuously and actively scans the institutional for regulated silences, voicing allegiance to lived collective experience through the embodied memory of a 'witness.'"[6] Many Kentucky burley farmers have co-opted institutional heritage discourses as a means of defending tobacco production, but there are differences between their usages and how this discourse continues to be used in spheres such as the news media. When used by farmers, "tobacco is our heritage" is used to refer to how they came to be who they are and do what they do through that which they have inherited: their land, their traditions (both tangible and intangible), their very identity. Unlike institutional heritage, vernacular heritage continues to claim that "we're still here" and that "tobacco pays my bills." Related to this, however, is the mantra that tobacco once paid the mortgage, put shoes on the children, and allowed many Kentuckians to attend college. This is a heritage-based argument but also an expression of nostalgia for a time when tobacco could do these things.

Tobacco nostalgia is in part about the loss of economic value, but such value is entangled with a sense of loss for a time when both tobacco and the tobacco man were respected. In this fluid Golden Age, the performance of a particular masculinity based on a mastery of traditional skills was highly valued in large part because the work was understood as central to the region's economy. Farmers make choices about what crops to grow for a number of reasons, including reasons that, although perhaps unarticulated, are tied up with traditionalized gendered meanings. Although tobacco production once depended on the labor of the entire family, and there are women who are heavily involved in its production, it is widely understood as a crop under male control. More than ever, perhaps, it is a crop produced primarily by men as women work off-farm jobs or diversify their farm operations. The cultural practices involved in the production of tobacco, when done with "pride," add up to the performance of a particular traditionalized masculinity, that of the tobacco man.

The changing symbolism of the crop combines with changing cultural practices to create nostalgia not only for past ways of production but for performed identities that are understood as in decline or even as no longer existing, because the performance of the mastery of particular skills—those of their fathers, grandfathers, and perhaps their younger selves—is no longer necessary or even possible. They do not want to go back to the work of those days because it would be economically unfeasible; instead, tobacco farmers express nostalgia for a time when tobacco was treated with respect by tobacco men, who themselves were treated with respect in their communities and who had a great deal of social and economic power. Being a burley tobacco man isn't equivalent, in terms of social status, monetary gain, or the performance of a regionally valued masculinity, with being a vegetable farmer. While it can be argued that some of the alternatives to tobacco that are currently being promoted have symbolic values with connections to identities and "ways of life"—such as developing vineyards or organic vegetable operations—they are so different from the performances and practices of the tobacco man that they do not offer a true substitution. Instead they require a paradigm shift.

Diversification efforts have failed to take into account that as farmers are being told to diversify, they are at the same time being told not only to abandon tradition—both intangible knowledge and tangible resources such as equipment and land—but to abandon a distinct performed masculinity built on the mastery of particular skills. These efforts are motivated in part by the changed symbolism; the state has an interest in creating a new narrative about Kentucky agriculture that is based on healthy farm products rather than tobacco. Such efforts are also gendered, as they depend on professionalizing what was once largely considered "women's work": vegetables, flowers, and value-added products. Because crops such as vegetables and flowers grown for local consumption have traditionally—but certainly not always—been raised primarily by women, such crops are understood to be "women's crops." Women have always played central roles in the production of tobacco, and they have had the primary responsibility for other farm activities and for the generation of particular kinds of farm income, however unacknowledged. As women take leading roles in situations in which tobacco production is indeed being "replaced," they are continuing and expanding upon women's traditional roles on farms—but often in much more public ways. While women are often leading the way in diversifying family farms, their current roles—like their historical roles—have gone widely unacknowledged. As tobacco farmers are asked to engage in vegetable and flower production, women's work continues to be devalued.

The key to successfully diversifying is to create or tap into a niche market—and to hope that your neighbors do not attempt to emulate your success and flood the market. In Kentucky, like the rest of the nation, there are increasing calls for agricultural production for local markets. From the very beginning of tobacco production by Europeans in colonial America, tobacco has been a global industry. While tobacco growers always knew that their product's economic value went beyond the local, the immediate symbolic value for them did not. Local value lay in the successful performance of a tobacco-man identity for other farmers through the production of a good crop of tobacco. Such local performance has lost much of its value as farmers do less physical labor, receiving stations replace auctions, and stigma has tarnished the symbolic value of the crop and the traditional practice of raising it. Meanwhile, tobacco farmers have become increasingly aware of their role in a global market, as manufacturers have bought more and more tobacco from overseas, and therefore farmers see themselves as being in competition with farmers in places such as Zimbabwe and Brazil whom they will never meet.[7] In addition, one reason that many tobacco farmers feel secure in the future of tobacco is the rising rate of smoking in China and other parts of Asia. So even though Americans are smoking less and less, farmers see the Chinese—whom they will also never meet—as potential customers. This also helps them to further distance their tobacco cultivation from the health effects of smoking.

Diversification rhetorics ask tobacco farmers to move from global to more local markets. In order to serve local markets, not only must tobacco farmers obtain new equipment, knowledge, and skills, but they must also learn to perform not for tobacco buyers and other farmers, but for consumers. While the purpose of this book is an examination of the implications for tobacco farming families of the changing contexts of this traditional occupation and "way of life," this case study has implications that reach beyond tobacco farming families. According to Eric Ramírez-Ferrero, writing in the aftermath of the farm crisis of the 1980s and '90s, "American agriculture is in a period of transition."[8] In this sense, then, tobacco farmers are not alone. Since the period of Ramírez-Ferrero's research, public dialogues about agriculture have been on the rise. The public discourses surrounding tobacco, through contradictory messages, reflect the transitions in American agriculture more generally, as some argue that only farmers willing to continually enlarge their operations will survive (reflecting a belief in the system of industrial agriculture that began in the early twentieth century) and others critique industrial agriculture in favor of smaller, more diverse farms. Sustainable agriculture advocates hope that American agriculture is on the precipice of real change, as consumers increasingly question not only the disappearance of small farms in their locales but also the use of petroleum products in moving farm products around the world; the application of chemicals that are injurious to the environment and to consumers; and the dangers of not knowing where their food comes from, as consumer scares about spinach, jalapeño peppers, green onions, and other vegetables tainted with E. coli and other bacteria sweep the nation.

In order for this movement to be successful, however, the challenges faced by all types of farmers who have long depended on a single crop must be more fully understood. Such farmers now farm as they do as a direct result of adhering to the "new agriculture" that developed early in the twentieth century, leading to increased dependence on monocrop farming. It is not as simple as telling farmers to "grow something else" that will replace tobacco (or other traditional farm products in other regions); certainly it is not as simple as telling tobacco farmers to grow vegetables for local people. As farmers consider their futures, they must grapple with issues of heritage, nostalgia, and traditionalized gendered performances of identity.

As I drove the back roads of the burley region of Kentucky throughout the harvest season of 2007, passing field after field of tobacco in various stages of cutting and housing, I thought about the many times I have heard comments about the landscape changing as a result of the disappearance of tobacco—comments that reflect the widespread perception that tobacco is gone from Kentucky. As if he's had similar thoughts, when I asked county agent Dan Grigson what he thought the level of public awareness was in terms of the amount of tobacco being raised, he said: "I think many people feel like tobacco is gone. You know, 'It's not around here anymore,' you know. If they'd be out in the country and drive around a little bit they still wouldn't see lots of acres of tobacco, because most of your tobacco is back off the road. I mean they'll see some driving down the highways, but I don't think the general public has any realization, [of] the amount of tobacco [that] is still being grown here." Despite the dramatically changed meanings of tobacco, tobacco farmers are still here. Despite the recategorization of tobacco as heritage, tobacco remains economically important to many Kentucky farmers.

Notes

Introduction

1. See Mary Hufford, *One Space, Many Places: Folklife and Land Use in New Jersey's Pinelands National Reserve* (Washington, DC: American Folklife Center, Library of Congress, 1986); Barbara Allen Bogart and Thomas J. Schlereth, *Sense of Place: American Regional Cultures* (Lexington: University Press of Kentucky, 1992).

2. Lucy R. Lippard, *The Lure of the Local: Senses of Place in a Multicentered Society*. (New York: New Press, 1998), 7.

3. Gregory Clark, *Rhetorical Landscapes in America: Variations on a Theme from Kenneth Burke* (Columbia: University of South Carolina Press, 2004), 9.

4. Will Snell, "Burley and Dark Tobaccos Issues and Outlook" (PowerPoint presentation at the Tobacco Merchants Association annual meeting and conference, Williamsburg, VA, May 19–21, 2008).

5. Chris Bickers, "Tobacco Growers in a Brave New World," *Southeast Farm Press* 32 (2005): 6.

6. The connections between the fields of folklore and rhetoric are lengthy and well-established, extending from our mutual claims to the influence of Gianbattista Vico, to the important connections between the work of Kenneth Burke and Dell Hymes, to current attempts to bring the fields together, such as this project. Roger Abrahams has argued that the function of folklore is at its base rhetorical; folklore texts and performances are intended to persuade. According to Abrahams, "Folklore, being traditional activity, argues traditionally; it uses argument and persuasive techniques developed in the past to cope with the recurrences of social problem situations." See Roger Abrahams, "Introductory Remarks to a Rhetorical Theory of Folklore." *Journal of American Folklore*, no. 81 (1968): 143–58. For a recent explication of the connections between folklore and rhetoric, past and potential, see Stephen Olbrys Gencarella, "Constituting Folklore: A Case for Critical Folklore Studies," *Journal of American Folklore* 122 (2009): 172–96. For folklore's connections to Vico, see Regina Bendix, *In Search of Authenticity* (Madison: University of Wisconsin Press, 1997), 28.

7. G. Clark, *Rhetorical Landscapes in America*, 9.

8. Ibid.

9. John Dorst, *The Written Suburb: An American Site, an Ethnographic Dilemma* (Philadelphia: University of Pennsylvania Press, 1989), 6.

10. As Henry Glassie has written, "Culture is not a problem with a solution. There are no conclusions. Studying people involves refining understanding, not achieving final proof. Perhaps if you observe people as though they were planets or orchids, proceeding without hypotheses is foolhardy, but that was not my intention." See Henry Glassie, *Passing the Time in Ballymenone: Culture and History of an Ulster Community* (Bloomington: Indiana University Press, 1982), 13.

11. Kenneth Burke, "Terministic Screens," in *Language as Symbolic Action: Essays on Life, Literature, and Method* (Berkeley: University of California Press, 1966), 45 (emphasis per original).

12. See, for instance, James Clifford and George E. Marcus, eds., *Writing Culture: The Poetics and Politics of Ethnography* (Berkeley: University of California Press, 1986).

13. Peter Benson has written about the past and present situation of tobacco production in North Carolina. See Peter Benson, *Tobacco Capitalism: Growers, Migrant Workers, and the Changing Face of a Global Industry* (Princeton: Princeton University Press, 2012), and "Good Clean Tobacco: Philip Morris, Biocapitalism, and the Social Course of Stigma in North Carolina," *American Ethnologist* 35 (2008): 357–79. Benson points out that North Carolina differs from other regions (*Tobacco Capitalism*, 31).

14. Will Snell, "The Buyout: Short-Run Observations and Implications for Kentucky's Tobacco Industry" (Lexington: Department of Agricultural Economics, University of Kentucky College of Agriculture, May 2005), http://www.ca.uky.edu/cmspubsclass/files/lpowers/policyoutlk/buyout_short-run.pdf.

15. It is for this reason that the title of this book implies that "Kentucky tobacco" is burley tobacco. My intention is not to diminish the importance of other types of tobacco or those who raise it; rather, it is to reflect the historical reality of burley's dominance both economically and culturally.

16. D. Wynne Wright, "Fields of Cultural Contradictions: Lessons from the Tobacco Patch," *Agriculture and Human Values* 22, no. 4 (2005): 467.

17. See Ann Ferrell, "'It's Really Hard to Tell the True Story of Tobacco': Stigma, Tellability, and Reflexive Scholarship," *Journal of Folklore Research* 49, no. 2 (2012): 127–52.

18. Roger Quarles serves as the president of the Burley Tobacco Growers Co-operative Association. However, in his interviews with me he stressed that he was speaking only for himself and not on behalf of the co-op.

19. Ray Cashman, "Critical Nostalgia and Material Culture in Northern Ireland," *Journal of American Folklore* 119 (2006): 154.

20. Erving Goffman, *Stigma: Notes on the Management of Spoiled Identity* (1963; rpt., New York: Touchstone, 1986).

21. Hayden White, *Metahistory: The Historical Imagination in Nineteenth-Century Europe* (Baltimore: Johns Hopkins University Press, 1973).

22. Of course, the history of the tobacco industry has also been approached through explicit critique, including analyses of the industry and the economic, medical, and cultural impact of tobacco worldwide. For instance, throughout his historical analysis of sugar production and consumption, Sidney Mintz makes comparisons between the spread of the consumption of sugar and of tobacco (as well as other "drug foods," including rum and tea). See Sidney Mintz, *Sweetness and Power: The Place of Sugar in Modern History* (New York: Viking-Penguin, 1985). Mintz argues that the "provision of low-cost food substitutes, such as tobacco, tea, and sugar, for the metropolitan laboring classes" served as a means of increasing labor output and that therefore "such substitutes figured importantly in balancing the accounts of capitalism" (148). For a recent critical history of the tobacco industry in the twentieth century, see Allan M. Brandt, *The Cigarette Century: The Rise, Fall, and Deadly Persistence of the Product That Defined America* (New York: Basic, 2007).

Here I am interested specifically in those histories that purport to present a narrative of tobacco in order to inform rather than critique. It is my hope that through my use of Hayden White's view of "history" as emplotted narrative, however, I make it clear that I am not arguing that these authors are not also interpreting. On the contrary, through their emplotment, the "histories" that I have chosen to include here either implicitly or explicitly argue that tobacco was a positive force in the development of the nation (and Kentucky).

Of particular note is W. F. Axton's 1975 publication *Tobacco and Kentucky* (Lexington: University Press of Kentucky, 1975) because it is the most commonly cited of those that I look to here. In his preface, this descendant of a Louisville-based tobacco company family (Axton-Fischer, purchased by Philip Morris in 1941) refers to his work as "the first full-length history of Kentucky's tobacco" (ix). *Tobacco and Kentucky* has become a default source for subsequent histories. However, although Axton was correct that there had been no volume that focused exclusively on telling the historical story of Kentucky tobacco until he took on the task, there were multiple volumes about the history of tobacco in America more broadly. Axton told much the same story as his predecessors, only he moved Kentucky to the center and of course updated the story.

23. Susan Wagner, *Cigarette Country: Tobacco in American History and Politics* (New York: Praeger, 1971), 8.

24. Axton, *Tobacco and Kentucky*, 3.

25. Ibid., 8.

26. Wagner, *Cigarette Country*, 7.

27. Iain Gately, *Tobacco: The Story of How Tobacco Seduced the World* (New York: Grove, 2001), 23.

28. Axton, *Tobacco and Kentucky,* 24; see also Gately, *Tobacco.*

29. Axton, *Tobacco and Kentucky,* 24–25.

30. Wagner, *Cigarette Country,* 14; Axton, *Tobacco and Kentucky,* 25.

31. Gately, *Tobacco,* 72.

32. Axton, *Tobacco and Kentucky,* 25.

33. Randall Elisha Greene, *The Leaf Sellers: A History of US Tobacco Warehouses: 1619 to the Present* (Lexington: Burley Auction Warehouse Association, 1996), 14.

34. John van Willigen and Susan C. Eastwood, *Tobacco Culture: Farming Kentucky's Burley Belt* (Lexington: University Press of Kentucky, 1998), 10.

35. Wagner, *Cigarette Country,* 25.

36. Axton, *Tobacco and Kentucky,* 44.

37. Wagner, *Cigarette Country,* 121. Tobacco adorns the columns in the Small Senate Rotunda, reconstructed in 1816 after the 1814 fire, as well as the twenty-eight columns lining the Hall of Columns, constructed in the mid-nineteenth century. Other agricultural representations include corncob capitals elsewhere in the Capitol. See Architect of the Capitol, "Architectural Features and Historic Spaces," http://www.aoc.gov/cc/architecture/index.cfm.

38. Axton, *Tobacco and Kentucky,* 44.

39. Van Willigen and Eastwood, *Tobacco Culture,* 3.

40. Wagner, *Cigarette Country,* 14.

41. John Ferdinand Dalziel Smyth, *Tour in the United States: The Present Situation, Population, Agriculture, Commerce, Customs, Manners and a Description of the Indian Nations* (1784); Whitefish, MT: Kessinger, 1968, 2: 127.

42. Axton, *Tobacco and Kentucky,* 30.

43. Marion B. Lucas, *A History of Blacks in Kentucky,* 2 vols. (Frankfort: Kentucky Historical Society, 1992), 1: xi.

44. See ibid.

45. Allan Kulikoff, *Tobacco and Slaves: The Development of Southern Cultures in the Chesapeake, 1680–1800* (Chapel Hill: University of North Carolina Press for the Institute of Early American History and Culture, 1986), 38.

46. Joseph C. Robert, *The Story of Tobacco in America* (New York: Alfred A. Knopf, 1949), 15.

47. Kulikoff, *Tobacco and Slaves,* 119.

48. Ann E. Kingsolver, "Farmers and Farmworkers: Two Centuries of Strategic Alterity in Kentucky's Tobacco Fields," *Critique of Anthropology* 27 (2007): 87–102. Kingsolver writes that she was taught as a child in her Kentucky school that there had been no slaves in the county in which she lived. Her research proved otherwise. Also see Pem Davidson Buck,

Worked to the Bone: Race, Class, and Privilege in Kentucky (New York: Monthly Review Press, 2001), on slavery in Kentucky and its relationship to the construction of race and class in the state.

49. Stephen A. Channing, *Kentucky: A Bicentennial History* (New York: W. W. Norton, 1977), 95.

50. Axton, *Tobacco and Kentucky*, 58.

51. Robert, *Story of Tobacco*, 103.

52. Ibid., 102.

53. Axton, *Tobacco and Kentucky*, 48.

54. John Morgan, "Dark-Fired Tobacco: The Origin, Migration, and Survival of a Colonial Agrarian Tradition," *Southern Folklore* 54 (1997): 145–84.

55. The USDA recognizes six major classes of US-grown tobacco: flue-cured, fire-cured, air-cured (of which burley is one type), cigar binder, cigar wrapper, and cigar filler. Within each class there are two or more different types, at one time totaling about twenty-six. See John Fraser Hart and Eugene Cotton Mather, "The Character of Tobacco Barns and Their Role in the Tobacco Economy in the United States," *Annals of the Association of American Geographers* 51 (1961): 288–93. Perique tobacco is of a separate, minor tobacco class.

56. Axton, *Tobacco and Kentucky*, 57.

57. Ibid., 71; Robert, *Story of Tobacco*, 120.

58. Axton, *Tobacco and Kentucky*, 62.

59. Ibid., 65.

60. Robert, *Story of Tobacco*, 116.

61. Ibid., 185–86. See also van Willigen and Eastwood, *Tobacco Culture*, 11–12; Axton, *Tobacco and Kentucky*, 69–70; and George Melvin Herndon, *William Tatham and the Culture of Tobacco* (Coral Gables: University of Miami Press, 1969), 409. I find it interesting that in spite of the consistent inclusion of this narrative in written histories, I did not encounter it in oral tradition, unlike other pieces of the metahistory given here. Flue-cured tobacco, too, has an origin narrative, in which a slave named Stephen accidentally discovered the benefits of curing bright tobacco with indirect heat (flues) rather than direct heat (cf. Greene, *Leaf Sellers*, 19).

62. Axton, *Tobacco and Kentucky*, 69.

63. Ibid.

64. Ibid., 70.

65. Robert, *Story of Tobacco*, 186.

66. Axton, *Tobacco and Kentucky*, 51.

67. Pete Daniel, *Breaking the Land: The Transformation of Cotton, Tobacco, and Rice Cultures since 1880* (Urbana: University of Illinois Press, 1986), 31.

68. Herndon, *William Tatham*, 435; Robert, *Story of Tobacco*, 68.

69. Herndon, *William Tatham,* 436.

70. Axton, *Tobacco and Kentucky,* 86.

71. Robert, *Story of Tobacco,* 200.

72. Ibid., 8.

73. William Tatham, *An Historical and Practical Essay on the Culture and Commerce of Tobacco,* reprinted in Herndon, *William Tatham,* 252.

74. Ibid., 148–78.

75. Robert, *Story of Tobacco,* 11–12.

76. Ibid., 79.

77. Axton, *Tobacco and Kentucky,* 82.

78. Robert, *Story of Tobacco,* 138.

79. Axton, *Tobacco and Kentucky,* 83.

80. A number of volumes chronicle this period. See John G. Miller, *The Black Patch War* (Chapel Hill: University of North Carolina Press, 1936); Christopher Waldrep, *Night Riders: Defending Community in the Black Patch, 1890–1915* (Durham: Duke University Press, 1993); Suzanne Marshall, *Violence in the Black Patch of Kentucky and Tennessee* (Columbia: University of Missouri Press, 1994); van Willigen and Eastwood, *Tobacco Culture,* 39–46; and Buck, *Worked to the Bone,* 104–15, as well as Robert Penn Warren's fictional account, *Night Rider* (Boston: Houghton Mifflin, 1939).

81. Axton, *Tobacco and Kentucky,* 91.

82. Van Willigen and Eastwood, *Tobacco Culture,* 41.

83. Wagner, *Cigarette Country,* 49.

84. Kentucky Department of Agriculture, *Kentucky Marketing Bulletin,* Frankfort, 1947, 7.

85. Robert, *Story of Tobacco,* 233.

86. Ibid., 201.

87. Axton, *Tobacco and Kentucky,* 105.

88. Van Willigen and Eastwood, *Tobacco Culture,* 50.

89. Axton, *Tobacco and Kentucky,* 109.

90. Robert, *Story of Tobacco,* 207–8.

91. Van Willigen and Eastwood, *Tobacco Culture,* 52.

92. Greene, *Leaf Sellers,* 56–57.

93. Burley Tobacco Growers Co-operative Association, *The Producer's Program: Fifty Golden Years and More* (Lexington: Burley Tobacco Growers Co-operative Association, 1991), vii.

94. Van Willigen and Eastwood, *Tobacco Culture,* 53; Robert, *Story of Tobacco,* 212.

95. Greene, *Leaf Sellers,* 56.

96. Axton, *Tobacco and Kentucky,* 116.

97. See, e.g., Daniel, *Breaking the Land,* 184.

98. Burley Tobacco Growers Co-operative Association, *Producer's Program,* 86; see also Daniel, *Breaking the Land,* 245.

99. There were three failed attempts in the late 1960s to move to poundage. The successful 1971 referendum gave farmers the choice of accepting the poundage system and quota cuts or losing support prices altogether (Kentucky Department of Agriculture, *Kentucky Agricultural News,* Frankfort, Apr. 1971, 1). Flue-cured had gone to poundage several years earlier.

100. Van Willigen and Eastwood, e.g., interpret it in this way; see *Tobacco Culture,* 61.

101. Kentucky Department of Agriculture, *Kentucky Department of Agriculture Bulletin,* Frankfort, Feb. 1966, 2.

102. William M. Snell, Laura Powers, and Greg Halich, "Tobacco Economies in the Post-Buyout Era," in *2008 Kentucky Tobacco Production Guide,* ed. Kenny Seebold (Lexington: Cooperative Extension Service, University of Kentucky College of Agriculture, 2008), 5.

103. A. Blake Brown, William M. Snell, and Kelly H. Tiller, "The Changing Political Environment for Tobacco—Implications for Southern Tobacco Farmers, Rural Economics, Taxpayers and Consumers" (paper presented at the Southern Agricultural Economics Association annual meeting, Memphis, TN, Feb. 2, 1999), 13.

104. Van Willigen and Eastwood, *Tobacco Culture,* 60.

105. Robert, *Story of Tobacco,* 106.

106. Ibid., 106–7.

107. Ibid., 170–71.

108. Ibid., 171.

109. Wagner, *Cigarette Country,* 68.

110. Robert, *Story of Tobacco,* 247–50.

111. Axton, *Tobacco and Kentucky,* 116.

112. Wagner, *Cigarette Country,* 78.

113. Axton, *Tobacco and Kentucky,* 125.

114. See Wagner, *Cigarette Country.*

115. Ibid., 130.

116. Ibid., 164.

117. Ibid., 217.

118. Axton, *Tobacco and Kentucky,* 120. Although Philip Morris began to make cigarettes in London in the 1850s (Gately, *Tobacco,* 185), the company did not become a major player in the US industry until the 1940s and '50s. The Marlboro brand first appeared in the 1920s, marketed as a woman's cigarette with the slogan "Mild as May" (ibid., 244). It was not until the mid-1950s that the brand was remade as a man's cigarette, with the Marlboro Man image becoming one of the most recognizable advertising campaigns in American history (ibid., 277–78).

119. Robert, *Story of Tobacco,* 247.

120. Wagner, *Cigarette Country,* 91.

121. Greene, *Leaf Sellers,* 92.

122. Gail Gibson, “Clinton: ‘We Can Do It,’” *Lexington Herald-Leader,* Apr. 10, 1998.

123. Ibid.

124. President’s Commission, “Tobacco at a Crossroad: A Call for Action,” Final Report of the President’s Commission on Improving Economic Opportunity in Communities Dependent on Tobacco Production while Protecting Public Health, May 14, 2001, http://govinfo.library.unt.edu/tobacco/FRFiles/FinalReport.htm.

125. Brandt, *Cigarette Century,* 432.

126. Ibid., 438.

127. Ibid., 434.

128. Ibid.; Benson, *Tobacco Capitalism,* 61–62.

129. Will Snell, “US Tobacco Grower Issues” (Lexington: Department of Agricultural Economics, University of Kentucky, Feb. 2003), http://www.uky.edu/Ag/TobaccoEcon/outlook.html (no longer available).

130. Tom Capehart, “US Tobacco Import Update 2005/6,” Electronic Outlook Report from the Economic Research Service, USDA, May 2007, www.ers.usda.gov. According to Capehart, “In the 17th and 18th centuries, North American producers provided all the tobacco consumed globally. Foreign production began in the 19th century, but the United States remained the major supplier through the mid-20th century. As cigarette consumption expanded in the early 20th century, imports of Oriental or ‘Turkish’ tobacco (a cigarette leaf not produced in the United States) began, but its use was negligible compared with the total. By the early 1960s, imports accounted for 10 percent of use and still consisted mostly of Oriental tobacco along with some cigar leaf. Imports rose to about 30 percent of total use in the 1970s and 1980s as flue-cured and burley arrivals gradually increased. The increase occurred because foreign tobacco, while of lesser quality, was cheaper than domestic leaf. Further, new cigarette manufacturing technologies enabled use of more lower quality leaf in meeting blending requirements” (3).

131. Will Snell, email communication with author, June 25, 2008.

132. Daniel, *Breaking the Land,* 267–68. The buyout of pool stocks in 1985 also included “large increases in no-net-cost assessments” (van Willigen and Eastwood, *Tobacco Culture,* 62). Despite attempts to ensure that tax dollars did not fund the tobacco program, a 1995 study found that a majority of Americans surveyed (55 percent) believed that taxpayers subsidized tobacco farmers; see David G. Altman, Douglas W. Levine, and George Howard, “Tobacco Farming and Public Health: Attitudes of the General Public and Farmers,” *Journal of Social Issues* 53 (1997): 113–28.

133. See Benson, *Tobacco Capitalism,* for a brief discussion of some of the politics around the final passage of a tobacco buyout. Also see Donald D. Stull, “Tobacco Is Going, Going . . . but Where?” *Culture and Agriculture*

32, no. 2 (2009): 54–72, for an overview of the buyout and its implications for western Kentucky burley growers.

134. Kelly Tiller, "Tobacco Buyout Top Ten," prepared for the University of Kentucky, North Carolina State University, and the University of Tennessee, Feb. 21, 2005, http://agpolicy.org/tobaccobuyout/tobuy/TopTen.pdf (no longer available).

135. The state of Maryland passed an actual tobacco buyout in 2000. Those who voluntarily agreed to be bought out sold their right to raise tobacco or to help others raise it.

136. Gately, *Tobacco,* 222.

137. Brandt, *Cigarette Century,* 391.

138. Also see Benson, "Good Clean Tobacco," for an analysis of Philip Morris's motives and the responses to them.

139. Ann E. Kingsolver, *Tobacco Town Futures: Global Encounters in Rural Kentucky* (Long Grove, IL: Waveland Press, 2012).

140. Hampton Henton, "FDA Would Help Farmers," op-ed, *Lexington Herald-Leader,* June 6, 2009.

141. Tom Capehart, "Trends in US Tobacco Farming," Electronic Outlook Report from the Economic Research Service, USDA, Nov. 2004, www.ers.usda.gov.

142. USDA, *2007 Census of Agriculture: United States Summary and State Data,* National Agriculture Statistics Service, Feb. 2009, http://www.agcensus.usda.gov/Publications/2007/Full_Report/index.asp.

143. Snell, "Burley and Dark Tobaccos."

144. Will Snell, "Outlook for Kentucky's Tobacco Industry," in *Agricultural Situation and Outlook* (Lexington: Cooperative Extension Service, University of Kentucky College of Agriculture, 2006).

145. USDA, *2007 Census of Agriculture: North Carolina State and County Data,* National Agriculture Statistics Service, Feb. 2009, http://www.agcensus.usda.gov/Publications/2007/Full_Report/Census_by_State/North_Carolina/index.asp.

146. Snell, "Burley and Dark Tobaccos."

147. Will Snell, "Census Data Reveal Significant and a Few Surprising Changes in Kentucky's Tobacco Industry," Department of Agricultural Economics, University of Kentucky College of Agriculture, Feb. 2009, http://www.ca.uky.edu/agecon/Index.php?p=259.

148. USDA, *2007 Census of Agriculture: Kentucky State and County Data,* National Agriculture Statistics Service, Feb. 2009, 27, http://www.agcensus.usda.gov/Publications/2007/Full_Report/Census_by_State/Kentucky/. It is important to note, however, that the drop from 30,000 to 8,113 farms is in part a drop on paper only, because it includes the loss of nonproducing quota owners resulting from the buyout. What would have been counted, for instance, as five separate farms with tobacco raised by a

single grower may now be a single farm with the same amount of tobacco (or more).

149. Snell, "Census Data Reveal."

150. USDA, *2007 Census of Agriculture: North Carolina State and County Data.* As I briefly touch on in part 1, flue-cured tobacco can be grown in larger acreages today because production has been mechanized in ways that burley has not.

Introduction to Part 1

1. See Lee Haring's discussion of the importance of context in the performance of folklore, in which he argues, "The items the interviewer receives and decodes are selected by the performer on the basis of their appropriateness to that audience at that moment." See Lee Haring, "Performing for the Interviewer: A Study of the Structure of Context," *Southern Folklore Quarterly* 36 (1972): 387. Although Haring is specifically discussing storytelling events, his argument is certainly applicable to Martin's (and other farmers') choices of both stories and other tobacco knowledge shared with me, as they deemed it appropriate.

1. Sowing the Seeds and Setting the Tobacco

1. Seed trays can be purchased to hold varying numbers of plants (ranging from 200 to 392), dependent on the thickness of the polystyrene and the size of the cell.

2. Snell, "Census Data Reveal."

3. Lu Ann Jones, *"Mama Learned Us to Work": Farm Women in the New South* (Chapel Hill: University of North Carolina Press, 2002), 5.

4. Of course, there can also be too much moisture. In contrast to 2007, the 2008 and 2009 crops were late getting set because of very wet springs.

5. Kentucky Climate Center at Western Kentucky University, "Fact Sheet: Historic Droughts in Kentucky, "http://www.kyclimate.org/factsheets/historicdroughts.html, accessed June 29, 2008.

6. See van Willigen and Eastwood, *Tobacco Culture,* 83–101, for a more detailed discussion of the evolution of tobacco setting technologies.

7. Eric Ramírez-Ferrero, *Troubled Fields: Men, Emotions, and the Crisis in American Farming* (New York: Columbia University Press, 2005), 112.

8. Daniel, *Breaking the Land,* 67.

9. Ibid., 263.

10. At the same time, however, MH residues have been a source of ongoing research and controversy. European leaf buyers have insisted on tobacco with less MH chemical residue, and the University of Kentucky

College of Agriculture has defended its use (it has passed muster with the Environmental Protection Agency) while also conducting research into the best ways to lower residues for the sake of the market. The university now recommends a combination of MH and other chemicals, as well as new spraying techniques. Some farmers question the safety of MH (in terms of their own health and the health of tobacco users) at the same time that they depend on it for the labor savings it provides them.

2. The Harvest through Preparation for Market

1. Daniel, *Breaking the Land,* 264.

2. This was apparently a practice introduced and promoted by the University of Kentucky during the twentieth century. See van Willigen and Eastwood, *Tobacco Culture,* 125.

3. According to George Duncan, who told me that farmers commonly overestimate the weight of a stick of tobacco, the maximum weight of a stick of green tobacco is about forty-eight pounds (email communication with author, May 22, 2009).

4. As used in Tatham, *Historical and Practical Essay.*

5. See Hart and Mather, "Character of Tobacco Barns."

6. Will Snell and Greg Halich, "Tobacco Economies in the Post-Buyout Era," in *2007 Kentucky Tobacco Production Guide,* ed. Kenny Seebold (Lexington: Cooperative Extension Service, University of Kentucky College of Agriculture, 2007).

7. Articles in the Kentucky Department of Agriculture newsletter in 1990 (Apr., Oct.) detail efforts both to recruit migrant labor and to help farmers hire such workers legally. One farmer quoted in the newsletter echoed what many farmers told me: "We're not trying to find cheap labor; we're just trying to guarantee a workforce" (Kentucky Department of Agriculture, *Agriculture—Kentucky's Pride.* Frankfort, Apr. 1990, 5). While the political rhetoric of recent decades locates blame with "illegal immigrants" for "taking" American jobs, in many cases such workers were and are actively recruited.

8. Research that thoroughly reflects the perspectives and experiences of Latino workers in tobacco was beyond the scope of this project, but it is a subject in need of attention. See Benson, *Tobacco Capitalism,* for discussion of the situation and the treatment of Latino workers in North Carolina.

9. Snell and Halich also attribute a growing shortage of migrant labor in part to "immigrant issues (e.g., border control of illegal labor, terrorism concerns)" ("Tobacco Economies," 8).

10. University of Kentucky College of Agriculture, "Weather and Management Changes Could Aid Tobacco Curing," press release, Sept. 7, 2007, http://www.ca.uky.edu/newsreleases/2007/Sep/weathertobacco.htm.

11. Minus the plastic, of course, such curing structures were used by colonial tobacco planters as well. According to Robert, "hanging the leaves on lines rather than letting them ferment in piles in the sun" began during the time of Rolfe, introduced by a colonist named Thomas Lambert (*Story of Tobacco,* 9).

12. For example, a 2006 story on National Public Radio by Kentucky native Noah Adams was entitled "Tobacco Barns: Stately Relics of a Bygone Era."

13. The term *bulk* is most often cited in written sources and is understood by agricultural professionals as the correct term (cf. van Willigen and Eastwood, *Tobacco Culture*). However, *book* and *bulk* are used by farmers to mean the same thing. Because I heard *book* most often, I chose to use it here.

14. Tatham, *Historical and Practical Essay,* 37.

15. According to Joseph C. Robert, the name *trash* was used for the bottom grade before the advent of the burley-blend cigarette, "but with the great popularity of the cigarette the once 'trash' developed into the most valuable part of the plant" (*Story of Tobacco,* 221). He was writing before this switched once again; tips are now the most valuable leaves and trash the least.

16. Kentucky Department of Agriculture, *Kentucky Agricultural News,* Frankfort, Jan. 1994, 2.

17. Tatham makes it clear that hands were tied in colonial times. Some have postulated that hand tying was learned from Native Americans. See Bennet D. Poage, ed., *Tobacco Church II: A Manual for Congregational Leaders* (Richmond: Kentucky Appalachian Ministry, Christian Church [Disciples of Christ] in Kentucky, 1995).

18. This process began in mid-nineteenth-century Virginia (Greene, *Leaf Sellers,* 18) and reached Lexington in 1904 (Axton, *Tobacco and Kentucky,* 86).

19. George Duncan, "An Overview of Baling Burley Tobacco: 1973–2002," PowerPoint presentation, University of Kentucky College of Agriculture, Mar. 2002, http://www.bae.uky.edu/ext/tobacco/Presentations/25Yrs_BalingBurley_files/frame.htm.

20. As quoted in van Willigen and Eastwood, *Tobacco Culture,* 158.

21. Greene, *Leaf Sellers,* 103.

22. See Benson, "Good Clean Tobacco."

3. Taking Tobacco to Market

1. Greene, *Leaf Sellers,* 53.

2. Van Willigen and Eastwood, *Tobacco Culture,* 40.

3. Axton, *Tobacco and Kentucky,* 105.

4. For instance, Eugene Baker Umberger Jr.'s 1975 MA thesis on

tobacco production, "Tobacco Farming: The Persistence of Tradition" (Center for Intercultural and Folk Studies, Western Kentucky University, 1975), cites a version of this story as printed in the *Courier Journal* in 1966 (77). Also see Robert, *Story of Tobacco,* 207.

5. Wendell Berry tells a story in which his father as a little boy witnesses his own father selling his tobacco and coming home "without a dime. They took it all. The crop . . . about paid the warehouse commission." His story ends there, although he adds an important coda: "My father saw men leave the warehouse crying and he said, when he was a little boy, 'If I can ever do anything about this, I'm going to.'" See Wendell Berry, quoted in Kimberly K. Smith, *Wendell Berry and the Agrarian Tradition: A Common Grace* (Lawrence: University Press of Kansas, 2003), 11. Wendell Berry's father, John Berry, went on to play a major role in the establishment of the Burley Tobacco Growers Co-operative Association. Berry's version is not told in joke form, but rather serves as an origin story for his family's legacy of involvement in both tobacco politics and (particularly in his case) advocacy for small farmers.

6. The date or even the decade of the beginning of the tobacco program is not something that most tobacco growers I have interviewed know, and so variants of all these stories are vaguely situated in time as "early in the century" or as "sometime" in the 1920s, '30s, or '40s. Interestingly, growers are similarly vague about the period in which the Night Riders were active. While this is clearly an example of the way that the tobacco narrative has been constructed differently in oral and written versions, it also might be understood in terms of Alessandro Portelli's argument that "false" memories have much to teach us about the meanings of events in people's lives and even what constitutes an "event." In an examination of a chronological shift in people's memories about the killing of a worker in Terni, Italy, Portelli argues that the date was remembered incorrectly because what traditional historiography might count as a single and isolated event may be only a portion of a series of "events" that come together to form a person's (or a community's) understanding of the event. See Alessandro Portelli, *The Death of Luigi Trastulli, and Other Stories: Form and Meaning in Oral History* (Albany: SUNY Press, 1991). It seems to me that events such as the activity of the Night Riders and the creation of the Burley Co-op and the tobacco program are similarly collapsed into one event and, further, that there is understood to be a cause-and-effect relationship: the program and the co-op were created because of the Black Patch Wars. Since in fact the "events" took place over a thirty-year period (the Black Patch Wars in 1906–1908, co-op formation in 1921, the program created through the Agricultural Adjustment Act first in 1933 but more permanently in 1938), knowledge of the actual dates only creates confusion; it makes more sense that they would

have happened during the same period. Of course, this situation differs from Portelli's in that these are historical events that the individuals I worked with did not experience firsthand because they were either not living or were too young to be fully aware of these events. So although this misremembering might best be understood through a lens of historical legend, Portelli's argument remains useful. I also encountered similar vagueness around the dates of more recent events, particularly the move from acreage to poundage and from hands to bales.

7. Melissa Walker has examined farmers' expressions of conflicting feelings about their dependence on government programs. See Melissa Walker, *Southern Farmers and Their Stories: Memory and Meaning in Oral History* (Lexington: University Press of Kentucky, 2006); Daniel (*Breaking the Land*); David B. Danbom (*Born in the Country: A History of Rural America* [Baltimore: Johns Hopkins University Press, 1995]); and others have described the contradictions in government farm policy beginning with the New Deal and the resulting conflicting impacts of government intervention on farmers.

8. Greene, *Leaf Sellers,* 116.

9. Farmers Tobacco Warehouse No. 1 was demolished in 2011. Jerry Rankin continued his auction business elsewhere.

10. I was treated with relative suspicion—although everyone was certainly friendly—by the employees and was told in clear terms what I could and could not photograph. In addition to visiting the Phillip Morris receiving station, I also accompanied the Waits family when they delivered a load of tobacco to the Burley Co-op.

11. According to Peter Benson, "Philip Morris hires graders from various regions to ensure against biased relationships with farmers" ("Good Clean Tobacco," 365).

Introduction to Part 2

1. Kenneth Burke, *A Rhetoric of Motives* (Berkeley: University of California Press, 1969), 41.

2. See Benson, *Tobacco Capitalism;* Brandt, *Cigarette Century.*

3. Brandt, *Cigarette Century,* 398.

4. Benson, *Tobacco Capitalism,* 40.

5. Ibid., 41.

6. Kentucky Department of Agriculture, *Kentucky Department of Agriculture Bulletin*, Frankfort, Apr. 1955, 1.

7. Kentucky Department of Agriculture, *Kentucky Agricultural News,* Frankfort, Jan. 2004, 4.

8. Jeffrey M. Duff and Martha L. Hall, comps., "Preliminary Inventory of the Records of the Department of Agriculture, Commonwealth of

Kentucky" (Frankfort: Division of Archives and Records, Department of Library and Archives, 1977), 2.

9. Ibid., 1.

10. Ibid., 2.

11. Because the name of the newsletter changed several times over the years, I will refer to it as "the KDA newsletter" here. See entries listed under Kentucky Department of Agriculture in the bibliography for a complete list of the names and the publication dates of issues cited in this book. The newsletter was published monthly, with occasional exceptions, through 1983, at which point it became a quarterly publication.

12. Dufford Hall, "Preliminary Inventory," 11.

13. See Deborah Fitzgerald, *Every Farm a Factory: The Industrial Ideal in American Agriculture* (New Haven: Yale University Press, 2003); Daniel, *Breaking the Land;* and Danbom, *Born in the Country.*

14. My use of *vernacular* can be understood through Margaret Lantis's definition of vernacular to mean "culture-as-it-is-lived appropriate to well-defined places and situations"; see Lantis, "Vernacular Culture," *American Anthropologist* 62, no. 2 (1960): 203.

15. Vernon Carstensen, "An Overview of American Agricultural History," in *Farmers, Bureaucrats, and Middlemen: Historical Perspectives on American Agriculture,* ed. Trudi Huskamp Peterson (Washington, DC: Howard University Press, 1980), 15.

16. Danbom, *Born in the Country,* 111–12.

17. Daniel, *Breaking the Land,* 16. Although the Extension Service was officially created by the USDA in 1914, it was "pioneered in Texas in 1902 by a scientist named Seaman Knapp" and quickly spread through southern states. With the Smith-Lever Act, federal matching funds were made available through the USDA "to help states expand extension programming" (Walker, *Southern Farmers and Their Stories,* 21).

18. Danbom, *Born in the Country,* 217.

19. Ibid., 235–38. These government programs worked against each other to the detriment of farmers in many ways, including the fact that while the Extension Service was promoting efficient farming methods in order to increase productivity, New Deal programs were being created that placed limits on production. See Danbom, *Born in the Country,* 213; Daniel, *Breaking the Land,* 245.

20. Women were directly appealed to through a long-standing pedagogical column that included recipes, advice for the hostess, and household tips. This section went through a series of names over the years, including "For the Ladies," "In the Home," and "Kookin' Korner," until it disappeared in the early 1970s. Since the female audience was so explicitly carved out through this space, we can presume that the rest of the newsletter was written primarily with a male audience in mind.

21. Burke, *Rhetoric of Motives,* 55 (emphasis per original).

22. This strategy can be understood as the use of what, in classical rhetoric, is called ceremonial or epideictic speech. According to Gregory Clark, "The traditional purpose of epideictic speech is to display collective values as enacted in exemplary stories of praiseworthy people—and in their image those addressed are invited to make themselves over" (*Rhetorical Landscapes in America,* 20).

23. For these terms I am indebted to Amy Shuman (numerous personal communications with the author), who suggests *self-conscious* and *self-evident* as terms signifying different attitudes toward particular traditions, following Barbara Kirshenblatt-Gimblett's discussion of heritage as one result of the loss of "self-evident" status; see Barbara Kirshenblatt-Gimblett, "Sounds of Sensibility," *Judaism* 47 (1998): 50. Rhetorically, self-conscious statements call attention to their production; self-evident statements rely on naturalized or assumed truths and conceal their production. Kirshenblatt-Gimblett's discussion is useful because it shows how self-conscious and self-evident discourses are always interdependent. Kirshenblatt-Gimblett demonstrates how "heritage" invents or points to prior, seemingly self-evident cultural practices.

24. Ibid., 52.

25. As quoted in David Brett, *The Construction of Heritage* (Cork, Ireland: Cork University Press, 1996), 46.

26. James F. Abrams, "Lost Frames of Reference: Sightings of History and Memory in Pennsylvania's Documentary Landscape," in *Conserving Culture: A New Discourse on Heritage,* ed. Mary Hufford (Urbana: University of Illinois Press, 1994), 25.

27. Kirshenblatt-Gimblett, "Sounds of Sensibility," 52.

28. Snell, "Burley and Dark Tobaccos."

29. In this analysis of the KDA newsletters I will only occasionally mention coverage of dark tobacco, when it is relevant. However, in instances where I have counted the number of articles about tobacco or where I discuss general trends in the coverage of tobacco, I have included all types of tobacco in my analysis.

30. See Ferrell, "It's Really Hard to Tell."

4. Tobacco's Move from Self-Evident to Self-Conscious Tradition

1. Wagner, *Cigarette Country,* 74.

2. Axton, *Tobacco and Kentucky,* 116.

3. Kentucky Department of Agriculture, *Kentucky Marketing Bulletin,* Frankfort, Jan. 1945, 1.

4. Ibid.

5. Kentucky Department of Agriculture, *Kentucky Marketing Bulletin,* Frankfort, Feb. 1945, 6.

6. Kentucky Department of Agriculture, *Kentucky Marketing Bulletin,* Frankfort, Jan. 1946, 2.

7. Kentucky Department of Agriculture, *Kentucky Marketing Bulletin,* Frankfort, July 1948, 2.

8. Brandt, *Cigarette Century,* 159.

9. Roger William Riis, "How Harmful Are Cigarettes?" *Reader's Digest,* Jan. 1950, 1–11.

10. Wagner, *Cigarette Country,* 78. *Reader's Digest* had begun tobacco research a decade before, however (see Brandt, *Cigarette Century,* 79).

11. See Benson, *Tobacco Capitalism,* for discussion of the development of the industry response in this period.

12. Brandt, *Cigarette Century,* 159.

13. Kentucky Department of Agriculture, *Kentucky Agriculture Bulletin,* Frankfort.

14. Kentucky Department of Agriculture, *Kentucky Agriculture Bulletin,* Frankfort, May 1950, 4.

15. These series serve as prime examples of visual epideictic speech. See the introduction to part 2, n. 22.

16. Kentucky Department of Agriculture, *Kentucky Department of Agriculture Bulletin,* Frankfort, Mar. 1955, 6.

17. Ibid.

18. Barbara Kirshenblatt-Gimblett, *Destination Culture: Tourism, Museums, and Heritage* (Berkeley: University of California Press, 1998), 7.

19. Leslie Prosterman, *Ordinary Life, Festival Days: Aesthetics in the Midwestern County Fair* (Washington, DC: Smithsonian Institution Press, 1995), 188.

20. Kentucky Department of Agriculture, *Kentucky Department of Agriculture Bulletin,* Frankfort, Apr. 1955, 1.

21. Benson, *Tobacco Capitalism,* 45.

22. Kentucky Department of Agriculture, *Kentucky Department of Agriculture Bulletin,* Frankfort, May 1956, 8. For discussion of the introduction of this manufacturing process, see Brandt, *Cigarette Century,* 359; and Axton, *Tobacco and Kentucky,* 126–27.

23. When brands such as Camel, Chesterfield, and Lucky Strike were introduced, the term *quality tobacco* was frequently used in marketing campaigns. Farmers occasionally were even pictured in ads for brands such as Lucky Strike and Camel. According to Benson, one strategy currently being used by Philip Morris in its effort to rebrand itself as concerned about public health is to tout the "quality" of the tobaccos that it uses in its products (*Tobacco Capitalism,* 151–52). It is interesting to consider that this is in fact a return to practices of the early twentieth century rather than a new strategy.

24. Kentucky Department of Agriculture, *Kentucky Department of Agriculture Bulletin,* Frankfort, May 1956, 7.

25. Ibid.

26. Ibid.

27. Kentucky Department of Agriculture, *Kentucky Department of Agriculture Bulletin,* Frankfort, Dec. 1959, 2.

28. Kentucky Department of Agriculture, *Kentucky Department of Agriculture Bulletin,* Frankfort, Dec. 1960, 1.

29. Kentucky Department of Agriculture, *Kentucky Department of Agriculture Bulletin,* Frankfort, Feb. 1961, 4.

30. Kentucky Department of Agriculture, *Kentucky Department of Agriculture Bulletin,* Frankfort, Nov. 1960, 4.

31. Kentucky Department of Agriculture, *Kentucky Department of Agriculture Bulletin,* Frankfort, July 1962, 1.

32. Ibid.

33. For an examination of 1930s tobacco queens, see Blain Roberts, "A New Cure for Brightleaf Tobacco: The Origins of the Tobacco Queen during the Great Depression," *Southern Cultures* (Summer 2006): 30–52.

34. Kentucky Department of Agriculture, *Kentucky Department of Agriculture Bulletin,* Frankfort, Feb. 1964, 1–2.

35. Ibid., 2.

36. Brandt, *Cigarette Century,* 160.

37. Ibid., 186.

38. Ibid.

39. Ibid., 183.

40. Kentucky Department of Agriculture, *Kentucky Department of Agriculture Bulletin,* Frankfort, Mar. 1964, 2.

41. Ibid.

42. Kentucky Department of Agriculture, *Kentucky Department of Agriculture Bulletin,* Frankfort, Mar. 1967, 1.

43. Kentucky Department of Agriculture, *Kentucky Department of Agriculture Bulletin,* Frankfort, Aug. 1967, 1.

44. Benson, *Tobacco Capitalism,* 144, 145.

45. Kentucky Department of Agriculture, *Kentucky Department of Agriculture Bulletin,* Frankfort, Aug. 1967, 2.

46. Kentucky Department of Agriculture, *Kentucky Department of Agriculture Bulletin,* Frankfort, Oct. 1967, 2.

47. Kentucky Department of Agriculture, *Kentucky Department of Agriculture Bulletin,* Frankfort, Feb. 1968, 2.

5. Tobacco under Attack

1. "A Silent Auction," *Kentucky Post,* editorial reprinted from the *Maysville-Ledger Independent,* Nov. 20, 2003.

2. Barbara Kirshenblatt-Gimblett, "Theorizing Heritage," *Ethnomusicology* 39 (1995): 369.

3. Kentucky Department of Agriculture, *Kentucky Agricultural News*, Frankfort, Oct. 1970, 1.

4. The Tobacco Institute (TI), established in 1958, served as the public relations arm of the tobacco industry until it closed following the 1998 Master Settlement Agreement. See Brandt, *Cigarette Century*, for a discussion of the history and tactics of the TI.

5. See Axton, *Tobacco and Kentucky;* Robert K. Heimann, *Tobacco and Americans: The Tobacco Custom in America from Early Colonial Times to the Present* (New York: McGraw Hill, 1960); and Wagner, *Cigarette Country.*

6. Kentucky Department of Agriculture, *Kentucky Agricultural News*, Frankfort, Aug. 1974, 6.

7. Ibid.

8. However, I have yet to see a visual depiction of tobacco work that includes the newer float beds and greenhouses in place of the old seed beds, and often the work in the field is being done with the aid of a horse or mule rather than a tractor: the period being honored by tobacco heritage is "the old days."

9. G. Clark, *Rhetorical Landscapes in America*, 9.

10. Kentucky Department of Agriculture, *Kentucky Agricultural News*, Frankfort, Nov. 1972, 6. There have historically been differing opinions on the role and usefulness of the Burley Tobacco Growers Co-operative Association, because following the establishment of the federal tobacco program (specifically in 1941) the co-op's primary purpose became managing the pool stocks; therefore, many felt it did very little in terms of finding new markets. Since the end of the program, the role of the co-op has been increasingly questioned, with one result an unsuccessful lawsuit arguing that the co-op should be disbanded and its assets (in the hundreds of millions of dollars) divided among its members (defined as anyone with a financial interest in the growing of burley tobacco). In recent years the co-op has worked to re-create itself as a cooperative marketing association, mostly through developing markets in China.

11. Kentucky Department of Agriculture, *Kentucky Agricultural News*, Frankfort, Feb. 1975.

12. Kentucky Department of Agriculture, *Kentucky Agricultural News*, Frankfort, Aug. 1975, 3.

13. Kentucky Department of Agriculture, *Kentucky Agricultural News*, Frankfort, Apr. 1975, 1.

14. Kentucky Department of Agriculture, *Kentucky Agricultural News*, Frankfort, Jan. 1976, 2.

15. Kentucky Department of Agriculture, *Kentucky Agricultural News*, Frankfort, Apr. 1976, 6.

16. Ibid.

17. Kentucky Department of Agriculture, *Kentucky Agricultural News*, Frankfort, Sept. 1976, 4.

18. Duncan Murrell, "The Duke," *Southern Cultures* 12 (2006): 9.

19. Kentucky Department of Agriculture, *Kentucky Agricultural News*, Frankfort, Nov. 1976, 2.

20. Ibid.

21. Brandt, *Cigarette Century*, 87.

22. Ibid., 5.

23. Ibid., 281.

24. Kentucky Department of Agriculture, *Kentucky Agricultural News*, Frankfort, Sept. 1977, 2.

25. Ibid.

26. See Benson, *Tobacco Capitalism*, 98–102, for a discussion of the costs of the tobacco program to American taxpayers even after the passage of the "no-net-cost" legislation.

27. Benson provides an important reading of this campaign, relying in part upon internal R. J. Reynolds documents to demonstrate that farmers and others within the industry were its intended audience (*Tobacco Capitalism*, 108–11).

28. Kentucky Department of Agriculture, *Kentucky Agricultural News*, Frankfort, Nov. 1978, 2.

29. Ibid.

30. Ibid., 3.

31. Ibid.

32. See Brandt, *Cigarette Century*, 307, for a discussion of the industry's increasing loss of credibility over the 1980s–1990s.

33. Kentucky Department of Agriculture, *Kentucky Agricultural News*, Frankfort, July 1979, 1.

34. Kentucky Department of Agriculture, *Kentucky Agricultural News*, Frankfort, Jan. 1980, 4.

35. Kentucky Department of Agriculture, *Kentucky Agricultural News*, Frankfort, Aug. 1981, 2.

36. Benson, *Tobacco Capitalism*, 137.

37. Kentucky Department of Agriculture, *Kentucky Agricultural News*, Frankfort, July 1984, 1.

38. Ibid., 3.

39. Kentucky Department of Agriculture, *Kentucky Agricultural News*, Frankfort, Aug. 1982, 5.

40. See Roberts, "New Cure for Brightleaf Tobacco."

41. Kentucky Department of Agriculture, *Kentucky Agricultural News*, Frankfort, Oct. 1984, 4.

42. Ibid.

43. Ibid.

44. Burke, *Rhetoric of Motives,* 55 (emphasis per original).

45. Tobacco Institute, "Kentucky's Tobacco Heritage," pamphlet (Washington, DC: Tobacco Institute, n.d.), 1.

46. Ibid., 16.

47. Benson, *Tobacco Capitalism,* 96. While there are a number of tobacco museums across the tobacco belt, there is not one in Kentucky. I was told by a former warehouseman that there was a collaborative effort by the major tobacco companies—including Reynolds—in the mid-1980s to establish a burley museum, but it never materialized.

48. Kentucky Department of Agriculture, *Kentucky Agricultural News.* Frankfort, Apr. 1985, 1.

49. Ibid., 2.

50. Kentucky Department of Agriculture, *Kentucky Agricultural News,* Frankfort, Oct. 1985, 10.

51. Kentucky Department of Agriculture, *Kentucky Agricultural News,* Frankfort, Oct. 1986, 2.

52. Ibid., 6.

53. Kentucky Department of Agriculture, *Agriculture—Kentucky's Pride,* Frankfort, Apr. 1989, 7.

54. Kentucky Department of Agriculture, *Agriculture—Kentucky's Pride,* Frankfort, Jan. 1990, 8.

55. Kentucky Department of Agriculture, *Agriculture—Kentucky's Pride,* Frankfort, Apr. 1990, 5.

56. Kentucky Department of Agriculture, *Agriculture—Kentucky's Pride,* Frankfort, July 1990, 4.

57. Ibid.

58. Kentucky Department of Agriculture, *Agriculture—Kentucky's Pride,* Frankfort, Oct. 1990, 3.

59. Kentucky Department of Agriculture, *Kentucky Agricultural News,* Frankfort, Jan. 1992, 3.

60. Burley Tobacco Growers Co-operative Association, *Producer's Program,* 19.

61. Greene, *Leaf Sellers,* 12.

62. Kentucky Department of Agriculture, *Kentucky Agricultural News,* Frankfort, Apr. 1993, 1.

63. Kentucky Department of Agriculture, *Kentucky Agricultural News,* Frankfort, Jan. 1994, 2.

64. Ibid.

65. For a discussion of the increase in nicotine levels, see Brandt, *Cigarette Century,* 359.

66. Ibid., 365.

67. Kentucky Department of Agriculture, *Kentucky Agricultural News,* Frankfort, Apr. 1994, 2.

68. Ibid.

69. Kentucky Department of Agriculture, *Kentucky Agricultural News*, Frankfort, July 1994, 3.

70. Ibid.

71. Kentucky Department of Agriculture, *Kentucky Agricultural News*, Frankfort, Jan. 1996, 1.

72. Brandt, *Cigarette Century*, 409; Brandt also provides an in-depth discussion of previous lawsuits.

73. Kentucky Department of Agriculture, *Kentucky Agricultural News*, Frankfort, Fall 1998, 4.

74. Kentucky Department of Agriculture, *Kentucky Agricultural News*, Frankfort, Apr. 2000, 1.

75. Ibid., 3.

76. Kentucky Department of Agriculture, *Kentucky Agricultural News*, Frankfort, July 2000, 8.

77. Ibid.

78. Ibid., 9.

79. Kentucky Department of Agriculture, *Kentucky Agricultural News*, Frankfort, Jan. 2001, 2.

80. Ibid., 4.

81. Kentucky Department of Agriculture, *Kentucky Agricultural News*, Frankfort, Oct. 2001, 1.

82. Kentucky Department of Agriculture, *Kentucky Agricultural News*, Frankfort, July 2003, 7.

83. Ibid. (my emphasis).

84. Ibid. (my emphasis).

85. Kentucky Department of Agriculture, *Kentucky Agricultural News*, Frankfort, Oct. 2003, 4.

86. Kentucky Department of Agriculture, *Kentucky Agricultural News*, Frankfort, Jan. 2004, 4.

87. Kentucky Department of Agriculture, *Kentucky Agricultural News*, Frankfort, July 2004, 3.

88. Tiller, "Tobacco Buyout Top Ten."

89. Kentucky Department of Agriculture, *Kentucky Agricultural News*, Frankfort, Jan. 2005, 4.

90. City of Maysville. "Maysville Floodwall Mural Project," http://www.cityofmaysville.com/tourism/floodwall%20murals.html, accessed Sept. 5, 2008.

91. The changed name of the Carrollton festival is based on Carroll County Tourism, "Calendar of Events," http://www.carrolltontourism.com/calendar.htm, accessed Aug. 26, 2008 (no longer available). Based on a photo in the Kentucky Historical Society collections of the first festival committee, this festival appears to have started in 1934 as the Carrollton

Tobacco Festival. See Kentucky Historical Society, "First Carrollton Tobacco Festival Committee," digital collection, Apr. 9, 2006, http://www.kyhistory.com/cdm/singleitem/collection/ORP/id/629/rec/1.

92. Many other Kentucky towns once held tobacco festivals, including Shelbyville and Georgetown. A 2004 article in the *Cincinnati Enquirer* described the disagreements swelling around the Ripley, Ohio, Tobacco Festival, noting that "there are questions about how much longer Ripley will hang on to that heritage" both because of shrinking government quotas (this was just months before the buyout) and because "the use of tobacco has never been more socially unacceptable." See Matt Leingang, "A Fest for Tobacco?" *Cincinnati Enquirer,* Aug. 26, 2004.

93. Jim Warren, "Farmers at the End of Tobacco Road—Many Don't Trust Contract System," *Lexington Herald-Leader,* May 25, 2005.

94. Amy Wilson, "One Mourner, but No Prayer for Tobacco, Burley Sale Becomes a 'Funeral' One Year after Quota Buyout," *Lexington Herald-Leader,* Dec. 12, 2005.

95. This article is rare in that a woman was chosen to represent tobacco farmers, as the category "tobacco farmer" is most often assumed to be filled by men. Such women challenge assumptions as exceptions to the rule.

96. Regarding "terministic screens," Kenneth Burke writes, "Not only does the nature of our terms affect the nature of our observations . . . *many of the 'observations' are but implications of the particular terminology in terms of which the observations are made*" ("Terministic Screens," 46 [emphasis per original]).

97. Abrams, "Lost Frames of Reference," 28.

98. Kentucky Department of Agriculture, *Kentucky Agricultural News,* Frankfort, Nov. 1976, 2.

99. G. Clark, *Rhetorical Landscapes in America,* 20.

Introduction to Part 3

1. "Stigma" has been most notably considered by Erving Goffman in his 1963 work *Stigma,* in which he defines stigma as "an attribute that is deeply discrediting" to a person's social identity (3). Our "social identity" is an identity assigned to us by others, based on the categories into which others put us. Goffman was most interested in those who are born into a stigmatized category and those who through some circumstance move into an existing stigmatized category later in life.

2. See Benson, *Tobacco Capitalism.*

3. Kingsolver, *Tobacco Town Futures,* 44.

4. Altman et al., "Tobacco Farming and Public Health," 119.

5. In their revision of Goffman's notion of stigma, Link and Phelan add the dimension of power: "Stigma is entirely dependent on social, economic,

and political power—it takes power to stigmatize." See Bruce G. Link and Jo C. Phelan, "Conceptualizing Stigma," *Annual Review of Sociology* 27 (2001): 376. Therefore, they argue, an absence of power is required in situations of stigma. According to Link and Phelan, the difference between groups that are labeled and stereotyped but not stigmatized (such as lawyers, politicians, and white people) and those groups that *are* stigmatized comes down to who does and who does not have power.

6. Bill Bishop, "It's a Dog of a Deal," *Lexington Herald-Leader*, Nov. 20, 1998.

7. Thomas D. Clark, *Agrarian Kentucky* (Lexington: University Press of Kentucky, 1977), ix.

8. Ibid., 130.

9. Van Willigen and Eastwood, *Tobacco Culture*, 180.

10. Goffman reminds us that there are members of stigmatized groups who either do not recognize or refuse to accept stigma: "He bears a stigma but does not seem to be impressed or repentant about doing so" (*Stigma*, 6). However, he suggests not only an acknowledgment of membership in a stigmatized category but some degree of agreement about the validity of the basis of the category, followed by shame. Goffman, in his concern for stigma "management," does not recognize the questioning, on the part of the stigmatized, of the very basis of a stigmatized category. Rather than the "shame" that Goffman suggests the stigmatized feel, some farmers express a loss of social and economic power—but also a continued sense of pride, the opposite of shame. Loss of pride is also expressed, but more often loss of pride is an accusation made about farmers more generally, not in terms of a stigmatized self. However, as I will discuss in the next chapter, the two are intricately connected.

11. Benson, *Tobacco Capitalism*, 136.

12. Of course, there most likely are tobacco farmers who subsidize their income by growing marijuana, but I did not encounter such situations firsthand.

6. "*Now* is the good old days"

1. Cashman, "Critical Nostalgia and Material Culture," 138. See Cashman's important discussion of scholarly disregard toward/contempt for nostalgia.

2. James Baker Hall and Wendell Berry, *Tobacco Harvest: An Elegy* (Lexington: University Press of Kentucky, 2004), 4.

3. USDA, *2007 Census of Agriculture: County Profile, Henry, Kentucky*, National Agriculture Statistics Service, Feb. 2009, http://www.agcensus.usda.gov/Publications/2007/Online_Highlights/County_Profiles/Kentucky/index.asp.

4. Snell and Halich, "Tobacco Economies."

5. Hall and Berry, *Tobacco Harvest,* 3.

6. Kingsolver, "Farmers and Farmworkers," 93.

7. Kentucky Department of Agriculture, *Kentucky Department of Agriculture Bulletin,* Frankfort, Aug. 1954, 7.

8. Hall and Berry, *Tobacco Harvest,* 1.

9. Dorst, *Written Suburb,* 130.

10. Stuart Tannock, "Nostalgia Critique," *Cultural Studies* 9 (1995): 457.

11. Hall and Berry, *Tobacco Harvest,* 4.

12. Ibid., 18.

13. According to the website of the James Baker Hall Archive, in 1973 Baker was an assistant professor of English at the University of Kentucky and was already winning prizes for his photography. See http://jamesbakerhall.com, accessed Jan. 18, 2012.

14. Fred Davis, *Yearning for Yesterday: A Sociology of Nostalgia* (New York: Free Press, 1979), 13.

15. For a discussion of the historical roles of women on burley farms in the Central Ohio River Valley, particularly the relationship between social class and women's work in tobacco, see Jeffrey A. Duvall, "Knowing about the Tobacco: Women, Burley, and Farming in the Central Ohio River Valley," *Register of the Kentucky Historical Society* 108, no. 4 (2010): 317–46. On the gendered consequences of the shift to commercial tobacco production, see Buck, *Worked to the Bone,* 83.

16. See Carolyn E. Sachs, *The Invisible Farmers: Women in Agricultural Production* (Totowa, NJ: Rowan and Allenheld, 1983); Jacqueline Jones, "'Tore Up and a-Movin': Perspectives on the Work of Black and Poor White Women in the Rural South, 1865–1940," in *Women and Farming: Changing Roles, Changing Structures,* ed. Wava G. Haney and Jane B. Knowles (Boulder: Westview, 1988), 20; and Ramírez-Ferrero, *Troubled Fields,* 106.

17. However, there are women tobacco farmers, and it is important that they not be rendered invisible. For instance, Alice Baesler, mentioned throughout this book, raises over three hundred acres of tobacco. Although her husband now works beside her, for many years she raised it on her own when he was a mayor and then a congressman. She was often named by extension agents and farmers when I asked about women tobacco farmers, not only because she raises so much tobacco but also because she is well known for her farm advocacy work. She is in many ways, however, one exception that proves the rule. Another important example is Mattie Mack of Meade County, one of the most outspoken advocates for Kentucky tobacco farmers in the 1990s (see Gibson, "Clinton"). As an African American woman, Mack challenged both gender and racial stereotypes of the typical Kentucky tobacco farmer. I met fewer than a handful of couples in

which the woman described herself as actively involved in the tobacco, but of course they represent others.

18. Kentucky Department of Agriculture, *Kentucky Marketing Bulletin*, Frankfort, Jan. 1946, 1.

19. Virgil S. Steed, *Kentucky Tobacco Patch* (Indianapolis: Bobbs Merrill, 1947), 43.

20. Wendell Berry, "The Problem of Tobacco," in *Sex, Economy, Freedom, and Community: Eight Essays* (New York: Pantheon Books, 1994), 54 (emphasis per original).

21. I've not encountered the term *tobacco women* in other contexts and think his usage of it may be unique to the context of this interview.

22. Ramírez-Ferrero, *Troubled Fields*, 119.

23. Timothy C. Lloyd and Patrick B. Mullen, *Lake Erie Fisherman: Work, Identity, and Tradition* (Urbana: University of Illinois Press, 1990), 80.

24. Ibid., 88.

25. Amy Shuman, *Other People's Stories: Entitlement Claims and the Critique of Empathy* (Urbana: University of Illinois Press, 2005), 62.

26. Ibid., 64.

27. Hall and Berry, *Tobacco Harvest*, 1.

28. Cashman, "Critical Nostalgia and Material Culture," 154.

29. Ibid., 138–39.

30. Ibid., 148.

31. Davis, *Yearning for Yesterday*, 8.

32. Shuman, *Other People's Stories*, 62.

33. Lloyd and Mullen, *Lake Erie Fisherman*, 86.

34. Benson, *Tobacco Capitalism*, 358.

35. Kentucky Department of Agriculture, *Kentucky Marketing Bulletin*, Frankfort, May 1947, 2.

36. This was particularly true when tobacco quotas were measured in acres rather than pounds because the more leaves a farmer could get from the field to the barn to the market, the more money he made.

37. Cashman, "Critical Nostalgia and Material Culture," 146.

38. As quoted in ibid., 138.

7. "Why can't they just grow something else?"

1. Andy Mead, "Harvesting a Memory: An Homage to Our State Heritage, Berry and Baker Hall Evoke What It Once Was Like to Gather a Tobacco Crop," review of *Tobacco Harvest: An Elegy*, by James Baker Hall and Wendell Berry, *Lexington Herald-Leader*, Sept. 19, 2004.

2. Warren, "Farmers at the End of Tobacco Road."

3. University of Kentucky College of Agriculture, "Kentucky Turns the Page on Tobacco," *Magazine* 6 (2005): 4–9.

4. Beverly Fortune, "Burley Is Just a Memory Now," *Lexington Herald-Leader,* July 6, 2008.

5. Susan Reigler, "Wine Is in Kentucky's Past," *Louisville Courier Journal,* Apr. 22, 2007.

6. Mary Hufford, "Reclaiming the Commons: Narratives of Progress, Preservation, and Ginseng," in *Culture, Environment, and Conservation in the Appalachian South,* ed. Benita J. Howell (Urbana: University of Illinois Press, 2002), 112.

7. See Wright, "Fields of Cultural Contradictions," for a discussion of challenges to diversification.

8. Earlier movements that serve as predecessors to today's movement include the Southern Agrarians (Twelve Southerners, *I'll Take My Stand: The South and the Agrarian Tradition* [1930; rpt., Baton Rouge: Louisiana State University Press, 1977]); early and mid-twentieth-century "back-to-the-land" movements (Helen Nearing and Scott Nearing, *Living the Good Life* [Harborside, ME: Social Science Institute, 1954]); and the writings of contemporary agrarian authors such as Wendell Berry (Berry, *The Unsettling of America: Culture and Agriculture* [New York: Avon Books, 1977]).

9. However, cigarette brands such as Kentucky's Best, made with (but not exclusively with) Kentucky tobacco, do attempt to join this movement.

10. Governor's Office of Agricultural Policy, "Planning for the Future," http://agpolicy.ky.gov/planning/index.shtml, accessed Jan. 9, 2009 (my emphasis). This text had been updated and edited slightly when I accessed it in January 2012. The key terms I note here remained in use at that time, however.

11. Janet Patton, "Tobacco Warehouses Denied Funds: Patton Opposes Plan to Provide Nearly $5 Million," *Lexington Herald-Leader,* Nov. 2, 2001.

12. Governor's Office of Agricultural Policy, "Agricultural Development Board," http://agpolicy.ky.gov/board/index.shtml, accessed Oct. 23, 2007.

13. "Wrong Fertilizer—Don't Use Settlement Money to Grow Tobacco," editorial, *Lexington Herald-Leader,* May 12, 2005.

14. "Don't Backslide—Tobacco Barns No Place for Settlement Money," editorial, *Lexington Herald-Leader,* July 25, 2006.

15. USDA, *Kentucky Agricultural Statistics and Annual Report 2006–2007* (Washington, DC: National Agriculture Statistics Service, 2007).

16. Laura Powers, ed., "The Kentucky Agricultural Economic Outlook for 2008," Lexington: University of Kentucky College of Agriculture Cooperative Extension Service, http://www.ca.uky.edu/cmspubsclass/files/esm/2008KYOutlook.pdf, accessed Aug. 2, 2009.

17. Kentucky Department of Agriculture, "Kentucky Proud, 1,300-Plus Strong, Now State's Official Farm Marketer," http://www.kyagr.com/kyproud/review1.html, accessed Dec. 9, 2008 (no longer available).

18. Office of Agricultural Marketing and Product Production, Division

of Value-Added Plant Production, Kentucky Department of Agriculture, "Grape and Wine Program," http://www.kyagr.com/marketing/plantmktg/grape.htm, accessed Nov. 13, 2008.

19. Néstor García Canclini, *Transforming Modernity: Popular Culture in Mexico*, trans. Lidia Lozano (Austin: University of Texas Press, 1993), 12.

20. Tiller, "Tobacco Buyout Top Ten."

21. Snell, "Buyout," 1.

22. It is important to note that she is describing her own experiences as a farmer who has led the way in Kentucky in the introduction of organic farming (including tobacco), aquaculture, and other alternative crops.

23. L. Jones, *"Mama Learned Us to Work,"* 17.

24. John Fraser Hart and Ennis L. Chestang, "Turmoil in Tobaccoland," *Geographical Review* 86 (1996): 553.

25. Ann E. Kingsolver provides several examples of failed diversification narratives about peppers (*Tobacco Town Futures*, 45).

26. Hall and Berry, *Tobacco Harvest*, 17.

27. David Brancaccio, narr., "Growing Local, Eating Local," *NOW*, Public Broadcasting Corporation, transcript, Nov. 2, 2007, http://www.pbs.org/now/transcript/344.html.

28. Dan Ben-Amos, "The Seven Strands of *Tradition:* Varieties in Its Meaning in American Folklore Studies," *Journal of Folklore Research* 21 (1984): 97–131.

29. For "polar terms" see Kenneth Burke, "Definition of Man," in *Language as Symbolic Action: Essays on Life, Literature, and Method* (Berkeley: University of California Press, 1966), 11.

30. James Bau Graves, *Cultural Democracy: The Arts, Community, and the Public Purpose* (Urbana: University of Illinois Press, 2004), 43.

31. Dell Hymes, "Folklore's Nature and the Sun's Myth," *Journal of American Folklore* 88 (1975): 345–69.

32. Richard Handler and Jocelyn Linnekin, "Tradition, Genuine or Spurious?" *Journal of American Folklore* 97 (1984): 273–90.

33. Henry Glassie, "Tradition," *Common Ground: Key Words for the Study of Expressive Culture*, special issue, *Journal of American Folklore* 108 (1995): 395.

34. Handler and Linnekin, "Tradition, Genuine or Spurious?" 273.

35. One farmer quoted by D. Wynne Wright explicitly articulates the tangle of the tangible and intangible aspects of tobacco as tradition. According to this farmer, "I raise tobacco for several reasons. One is tradition. We come from a tobacco growing family. It's what my ancestors have done for generations." He goes on, "When I say tobacco is part of my tradition, I mean I have all the equipment and I know what I am doing, the plant itself is durable, and the market is certain" (Wright, "Fields of Cultural Contradictions," 472).

36. See Hymes, "Folklore's Nature," for the concept of traditionalization, or tradition as process rather than as merely rooted in time.

37. Of course, farming itself increasingly requires the skills of multiple occupations—from mechanic to lay veterinarian to agronomist.

38. Benson, *Tobacco Capitalism,* 364.

39. In Kentucky and in the United States as a whole, the number of women farmers is rising as the number of male farmers drops. In Kentucky the percentage of women farmers rose from 9.6 percent in 2002 to 10.7 in 2007. See USDA, *2007 Census of Agriculture: Kentucky State and County Data.*

40. L. E. Garkovich and J. Bokemeier, "Agricultural Mechanization and American Farm Women's Economic Roles," in *Women and Farming: Changing Roles, Changing Structure,* ed. Wava G. Hanley and Jane B. Knowles (Boulder: Westview Press, 1988), 216.

41. See Sachs, *Invisible Farmers.*

42. For critiques of the practical application, despite acknowledged cultural constructions, of "men's crops" and "women's crops," see Cheryl R. Doss, "Men's Crops? Women's Crops? The Gender Patterns of Cropping in Ghana," *World Development* 30 (2002): 1987–2000; Edward R. Carr, "Men's Crops and Women's Crops: The Importance of Gender to the Understanding of Agricultural and Development Outcomes in Ghana's Central Region," *World Development* 36, no. 5 (2008): 900–915.

43. For further discussion of the "tobacco-man" identity as it relates to agricultural diversification, see Ann K. Ferrell, "*Doing* Masculinity: Gendered Challenges to Replacing Burley Tobacco in Central Kentucky," *Agriculture and Human Values* 29, no. 2 (2012): 137–49.

Conclusion

1. Gately, *Tobacco,* 3.

2. Glassie, "Tradition," 395.

3. Jim Warren, "Farmers Kicking Tobacco Habit—Success of Alternative Crops Eases Burley's Hold on Region." *Lexington Herald-Leader* July 29, 2002.

4. Kirshenblatt-Gimblett, "Sounds of Sensibility," 7.

5. Kirshenblatt-Gimblett, "Theorizing Heritage," 369.

6. Abrams, "Lost Frames of Reference," 29.

7. See Kingsolver, *Tobacco Town Futures,* for a discussion of one tobacco community as it has coped with globalization.

8. Ramírez-Ferrero, *Troubled Fields,* 120.

Bibliography

Books, Articles, and Online Resources

Abrahams, Roger. "Introductory Remarks to a Rhetorical Theory of Folklore." *Journal of American Folklore*, no. 81 (1968): 143–58.

Abrams, James F. "Lost Frames of Reference: Sightings of History and Memory in Pennsylvania's Documentary Landscape." In *Conserving Culture: A New Discourse on Heritage*, edited by Mary Hufford, 24–38. Urbana: University of Illinois Press, 1994.

Adams, Noah. "Tobacco Barns: Stately Relics of a Bygone Era." *Morning Edition*. National Public Radio, November 28, 2006. http://www.npr.org/templates/story/story.php?storyId=6536351.

Altman, David G., Douglas W. Levine, and George Howard. "Tobacco Farming and Public Health: Attitudes of the General Public and Farmers." *Journal of Social Issues* 53 (1997): 113–28.

American Tobacco Company. *The American Tobacco Story*. American Tobacco Company, 1960.

Axton, W. F. *Tobacco and Kentucky*. Lexington: University Press of Kentucky, 1975.

Bailey, Phyllis L., ed. *Shelby County Tobacco Farmers: A Pictorial History*. Bagdad, KY: PB Books, 2007.

Ben-Amos, Dan. "The Seven Strands of *Tradition:* Varieties in Its Meaning in American Folklore Studies." *Journal of Folklore Research* 21 (1984): 97–131.

Bendix, Regina. *In Search of Authenticity*. Madison: University of Wisconsin Press, 1997.

Benson, Peter. "Good Clean Tobacco: Philip Morris, Biocapitalism, and the Social Course of Stigma in North Carolina." *American Ethnologist* 35 (2008): 357–79.

———. *Tobacco Capitalism: Growers, Migrant Workers, and the Changing Face of a Global Industry*. Princeton: Princeton University Press, 2012.

Berry, Wendell. "The Problem of Tobacco." In *Sex, Economy, Freedom, and Community: Eight Essays*, 53–68. New York: Pantheon Books, 1994.

———. *The Unsettling of America: Culture and Agriculture*. New York: Avon Books, 1977.

Bickers, Chris. "Tobacco Growers in a Brave New World." *Southeast Farm Press* 32 (2005): 6.

Bogart, Barbara Allen, and Thomas J. Schlereth. *Sense of Place: American Regional Cultures.* Lexington: University Press of Kentucky, 1992.

Brancoccio, David, narr. "Growing Local, Eating Local." *NOW.* Public Broadcasting Corporation, transcript, November 2, 2007. http://www.pbs.org/now/transcript/344.html.

Brandt, Allan M. *The Cigarette Century: The Rise, Fall, and Deadly Persistence of the Product That Defined America.* New York: Basic Books, 2007.

Brett, David. *The Construction of Heritage.* Cork, Ireland: Cork University Press, 1996.

Brown, A. Blake, William M. Snell, and Kelly H. Tiller. "The Changing Political Environment for Tobacco—Implications for Southern Tobacco Farmers, Rural Economies, Taxpayers and Consumers." Paper presented at the Southern Agricultural Economics Association annual meeting, Memphis, TN, February 2, 1999.

Buck, Pem Davidson. *Worked to the Bone: Race, Class, and Privilege in Kentucky.* New York: Monthly Review Press, 2001.

Burke, Kenneth. "Definition of Man." In *Language as Symbolic Action: Essays on Life, Literature, and Method,* 3–24. Berkeley: University of California Press, 1966.

———. *A Rhetoric of Motives.* Berkeley: University of California Press, 1969.

———. "Terministic Screens." In *Language as Symbolic Action: Essays on Life, Literature, and Method,* 44–62. Berkeley: University of California Press, 1966.

Burley Tobacco Growers Co-operative Association. *The Producer's Program: Fifty Golden Years and More.* Lexington: Burley Tobacco Growers Co-operative Association, 1991.

Canclini, Néstor García. *Transforming Modernity: Popular Culture in Mexico.* Translated by Lidia Lozano. Austin: University of Texas Press, 1993.

Capehart, Tom. "Trends in US Tobacco Farming." Electronic Outlook Report from the Economic Research Service, United States Department of Agriculture, November 2004. www.ers.usda.gov.

———. "US Tobacco Import Update 2005/06." Electronic Outlook Report from the Economic Research Service, United States Department of Agriculture, May 2007. www.ers.usda.gov.

Carr, Edward R. "Men's Crops and Women's Crops: The Importance of Gender to the Understanding of Agricultural and Development Outcomes in Ghana's Central Region." *World Development* 36, no. 5 (2008): 900–915.

Carstensen, Vernon. "An Overview of American Agricultural History."

In *Farmers, Bureaucrats, and Middlemen: Historical Perspectives on American Agriculture,* edited by Trudy Huskamp Peterson, 8–23. Washington, DC: Howard University Press, 1980.

Cashman, Ray. "Critical Nostalgia and Material Culture in Northern Ireland." *Journal of American Folklore* 119 (2006): 137–60.

Channing, Steven A. *Kentucky: A Bicentennial History.* New York: W. W. Norton, 1977.

Clark, Gregory. *Rhetorical Landscapes in America: Variations on a Theme from Kenneth Burke.* Columbia: University of South Carolina Press, 2004.

Clark, Thomas D. *Agrarian Kentucky.* Lexington: University Press of Kentucky, 1977.

Clifford, James, and George E. Marcus, eds. *Writing Culture: The Poetics and Politics of Ethnography.* Berkeley: University of California Press, 1986.

Danbom, David B. *Born in the Country: A History of Rural America.* Baltimore: Johns Hopkins University Press, 1995.

Daniel, Pete. *Breaking the Land: The Transformation of Cotton, Tobacco, and Rice Cultures since 1880.* Urbana: University of Illinois Press, 1986.

Davis, Fred. *Yearning for Yesterday: A Sociology of Nostalgia.* New York: Free Press, 1979.

Dorst, John. *The Written Suburb: An American Site, an Ethnographic Dilemma.* Philadelphia: University of Pennsylvania Press, 1989.

Doss, Cheryl R. "Men's Crops? Women's Crops? The Gender Patterns of Cropping in Ghana." *World Development* 30 (2002): 1987–2000.

Duff, Jeffrey M., and Martha L. Hall, comps. "Preliminary Inventory of the Records of the Department of Agriculture, Commonwealth of Kentucky." Frankfort: Division of Archives and Records, Department of Library and Archives, 1977.

Duncan, George. "An Overview of Baling Burley Tobacco: 1973–2002." PowerPoint presentation. University of Kentucky College of Agriculture. March 2002. http://www.bae.uky.edu/ext/tobacco/Presentations/25Yrs_BalingBurley_files/frame.htm.

Duvall, Jeffrey A. "Knowing about the Tobacco: Women, Burley, and Farming in the Central Ohio River Valley." *Register of the Kentucky Historical Society* 108, no. 4 (2010): 317–46.

Ferrell, Ann K. "*Doing* Masculinity: Gendered Challenges to Replacing Burley Tobacco in Central Kentucky." *Agriculture and Human Values* 29, no. 2 (2012): 137–49.

———. "'It's Really Hard to Tell the True Story of Tobacco': Stigma, Tellability, and Reflexive Scholarship." *Journal of Folklore Research* 49, no. 2 (2012): 127–52.

Fitzgerald, Deborah. *Every Farm a Factory: The Industrial Ideal in American Agriculture.* New Haven: Yale University Press, 2003.

Garkovich, L. E., and J. Bokemeier. "Agricultural Mechanization and American Farm Women's Economic Roles." In *Women and Farming: Changing Roles, Changing Structure,* edited by Wava G. Haney and Jane B. Knowles, 211–29. Boulder: Westview Press, 1988.

Gately, Iain. *Tobacco: The Story of How Tobacco Seduced the World.* New York: Grove, 2001.

Gencarella, Stephen Olbrys. "Constituting Folklore: A Case for Critical Folklore Studies." *Journal of American Folklore* 122 (2009): 172–96.

Glassie, Henry. *Passing the Time in Ballymenone: Culture and History of an Ulster Community.* Bloomington: Indiana University Press, 1982.

———. "Tradition." *Common Ground: Key Words for the Study of Expressive Culture,* special issue, *Journal of American Folklore* 108 (1995): 395–412.

Goffman, Erving. *Stigma: Notes on the Management of Spoiled Identity.* 1963. Reprint, New York: Touchstone, 1986.

Graves, James Bau. *Cultural Democracy: The Arts, Community, and the Public Purpose.* Urbana: University of Illinois Press, 2004.

Greene, Randall Elisha. *The Leaf Sellers: A History of US Tobacco Warehouses: 1619 to the Present.* Lexington: Burley Auction Warehouse Association, 1996.

Hall, James Baker, and Wendell Berry. *Tobacco Harvest: An Elegy.* Lexington: University Press of Kentucky, 2004.

Handler, Richard, and Jocelyn Linnekin. "Tradition, Genuine or Spurious?" *Journal of American Folklore* 97 (1984): 273–90.

Haring, Lee. "Performing for the Interviewer: A Study of the Structure of Context." *Southern Folklore Quarterly* 36 (1972): 383–98.

Hart, John Fraser, and Ennis L. Chestang. "Turmoil in Tobaccoland." *Geographical Review* 86 (1996): 550–72.

Hart, John Fraser, and Eugene Cotton Mather. "The Character of Tobacco Barns and Their Role in the Tobacco Economy in the United States." *Annals of the Association of American Geographers* 51 (1961): 288–93.

Heimann, Robert K. *Tobacco and Americans: The Tobacco Custom in America from Early Colonial Times to the Present.* New York: McGraw Hill, 1960.

Herndon, George Melvin. *William Tatham and the Culture of Tobacco.* Coral Gables: University of Miami Press, 1969.

Hufford, Mary. *One Space, Many Places: Folklife and Land Use in New Jersey's Pinelands National Reserve.* Washington, DC: American Folklife Center, Library of Congress, 1986.

———. "Reclaiming the Commons: Narratives of Progress, Preservation, and Ginseng." In *Culture, Environment, and Conservation in the Appalachian South,* edited by Benita J. Howell, 100–120. Urbana: University of Illinois Press, 2002.

Hymes, Dell. "Folklore's Nature and the Sun's Myth." *Journal of American Folklore* 88 (1975): 345–69.

James, I. *A Counter-Blaste to Tobacco.* London: Rodale Press, 1954.

Jones, Jacqueline. "'Tore Up and a-Movin': Perspectives on the Work of Black and Poor White Women in the Rural South, 1865–1940." In *Women and Farming: Changing Roles, Changing Structures,* edited by Wava G. Haney and Jane B. Knowles, 15–34. Boulder: Westview Press, 1988.

Jones, Lu Ann. *"Mama Learned Us to Work": Farm Women in the New South.* Chapel Hill: University of North Carolina Press, 2002.

Kentucky Climate Center at Western Kentucky University. "Fact Sheet: Historic Droughts in Kentucky." http://www.kyclimate.org/factsheets/historicdroughts.html. Accessed June 29, 2008.

Kentucky Department of Agriculture. *Agriculture—Kentucky's Pride.* Frankfort, 1988–1991.

———. *Kentucky Agricultural News.* Frankfort, 1969–1987, 1992–2008.

———. *Kentucky Agriculture Bulletin.* Frankfort, 1948–1954.

———. *Kentucky Department of Agriculture Bulletin.* Frankfort, 1954–1968.

———. *Kentucky Marketing Bulletin.* Frankfort, 1941–1948.

———. "Kentucky Proud, 1,300-Plus Strong, Now State's Official Farm Marketer." http://www.kyagr.com/kyproud/review1.html. Accessed December 9, 2008.

Kentucky Historical Society. "First Carrollton Tobacco Festival Committee." Digital collection. http://www.kyhistory.com/cdm/singleitem/collection/ORP/id/629/rec/1. Accessed September 5, 2008.

Kingsolver, Ann E. "Farmers and Farmworkers: Two Centuries of Strategic Alterity in Kentucky's Tobacco Fields." *Critique of Anthropology* 27 (2007): 87–102.

———. *Tobacco Town Futures: Global Encounters in Rural Kentucky.* Long Grove, IL: Waveland Press, 2012.

Kirshenblatt-Gimblett, Barbara. *Destination Culture: Tourism, Museums, and Heritage.* Berkeley: University of California Press, 1998.

———. "Sounds of Sensibility." *Judaism* 47 (1998): 49–78.

———. "Theorizing Heritage." *Ethnomusicology* 39 (1995): 367–80.

Kulikoff, Allan. *Tobacco and Slaves: The Development of Southern Cultures in the Chesapeake, 1680–1800.* Chapel Hill: University of North Carolina Press for the Institute of Early American History and Culture, 1986.

Lantis, Margaret. "Vernacular Culture." *American Anthropologist* 62, no. 2 (1960): 202–16.

Link, Bruce G., and Jo C. Phelan. "Conceptualizing Stigma." *Annual Review of Sociology* 27 (2001): 363–85.

Lippard, Lucy R. *The Lure of the Local: Senses of Place in a Multicentered Society.* New York: New Press, 1998.

Lloyd, Timothy C., and Patrick B. Mullen. *Lake Erie Fisherman: Work, Identity, and Tradition.* Urbana: University of Illinois Press, 1990.

Lucas, Marion B. *A History of Blacks in Kentucky.* 2 volumes. Frankfort: Kentucky Historical Society, 1992.

Marshall, Suzanne. *Violence in the Black Patch of Kentucky and Tennessee.* Columbia: University of Missouri Press, 1994.

Miller, John G. *The Black Patch War.* Chapel Hill: University of North Carolina Press, 1936.

Mintz, Sidney. *Sweetness and Power: The Place of Sugar in Modern History.* New York: Viking-Penguin, 1985.

Morgan, John. "Dark-Fired Tobacco: The Origin, Migration, and Survival of a Colonial Agrarian Tradition." *Southern Folklore* 54 (1997): 145–84.

Murrell, Duncan. "The Duke." *Southern Cultures* 12 (2006): 6–29.

Nearing, Helen, and Scott Nearing. *Living the Good Life.* Harborside, ME: Social Science Institute, 1954.

Poage, Bennett D., ed. *Tobacco Church II: A Manual for Congregational Leaders.* Richmond: Kentucky Appalachian Ministry, Christian Church (Disciples of Christ) in Kentucky, 1995.

Portelli, Alessandro. *The Death of Luigi Trastulli, and Other Stories: Form and Meaning in Oral History.* Albany: SUNY Press, 1991.

Powers, Laura, ed. "The Kentucky Agricultural Economic Outlook for 2008." Lexington: University of Kentucky College of Agriculture Cooperative Extension Service. http://www.ca.uky.edu/cmspubsclass/files/esm/2008KYOutlook.pdf. Accessed August 2, 2009.

President's Commission. "Tobacco at a Crossroad: A Call for Action." Final Report of the President's Commission on Improving Economic Opportunity in Communities Dependent on Tobacco Production while Protecting Public Health. May 14, 2001. http://govinfo.library.unt.edu/tobacco/FRFiles/FinalReport.htm.

Prosterman, Leslie. *Ordinary Life, Festival Days: Aesthetics in the Midwestern County Fair.* Washington, DC: Smithsonian Institution Press, 1995.

Ramírez-Ferrero, Eric. *Troubled Fields: Men, Emotions, and the Crisis in American Farming.* New York: Columbia University Press, 2005.

Riis, Roger William. "How Harmful Are Cigarettes?" *Reader's Digest,* January 1950, 1–11.

Robert, Joseph C. *The Story of Tobacco in America.* New York: Alfred A. Knopf, 1949.

Roberts, Blain. "A New Cure for Brightleaf Tobacco: The Origins of the Tobacco Queen during the Great Depression." *Southern Cultures* (Summer 2006): 30–52.

Sachs, Carolyn E. *The Invisible Farmers: Women in Agricultural Production.* Totowa, NJ: Rowan and Allanheld, 1983.

Shuman, Amy. *Other People's Stories: Entitlement Claims and the Critique of Empathy.* Urbana: University of Illinois Press, 2005.

Smith, Kimberly K. *Wendell Berry and the Agrarian Tradition: A Common Grace.* Lawrence: University Press of Kansas, 2003.

Smyth, John Ferdinand Dalziel. *Tour in the United States; The Present Situation, Population, Agriculture, Commerce, Customs, Manners and a Description of the Indian Nations.* Volume 2. 1784. Whitefish, MT: Kessinger, 1968.

Snell, Will. "Burley and Dark Tobaccos Issues and Outlook." PowerPoint presentation at the Tobacco Merchants Association annual meeting and conference, Williamsburg, VA, May 19–21, 2008.

———. "The Buyout: Short-Run Observations and Implications for Kentucky's Tobacco Industry." Lexington: Department of Agricultural Economics, University of Kentucky College of Agriculture, May 2005. http://www.ca.uky.edu/cmspubsclass/files/lpowers/policyoutlk/buyout_short-run.pdf.

———. "Census Data Reveal Significant and a Few Surprising Changes in Kentucky's Tobacco Industry." Department of Agricultural Economics, University of Kentucky College of Agriculture, February 2009. http://www.ca.uky.edu/agecon/Index.php?p=259.

———. "Outlook for Kentucky's Tobacco Industry." In *Agricultural Situation and Outlook,* 12–13. Lexington: Cooperative Extension Service, University of Kentucky College of Agriculture, 2006.

———. "US Tobacco Grower Issues." Lexington: Department of Agricultural Economics, University of Kentucky, February 2003. http://www.uky.edu/Ag/TobaccoEcon/outlook.html (no longer available).

Snell, Will, and Greg Halich. "Tobacco Economies in the Post-Buyout Era." In *2007 Kentucky Tobacco Production Guide,* edited by Kenny Seebold, 6–8. Lexington: Cooperative Extension Service, University of Kentucky College of Agriculture, 2007.

Snell, Will, Laura Powers, and Greg Halich. "Tobacco Economies in the Post-Buyout Era." In *2008 Kentucky Tobacco Production Guide,* edited by Kenny Seebold, 4–6. Lexington: Cooperative Extension Service, University of Kentucky College of Agriculture, 2008.

Steed, Virgil S. *Kentucky Tobacco Patch.* Indianapolis: Bobbs-Merrill, 1947.

Stull, Donald D. "Tobacco Is Going, Going . . . but Where?" *Culture and Agriculture* 32, no. 2 (2009): 54–72.

Sullivan, C. W., III. "Tobacco." In *Rooted in America: Foodlore of Popular Fruits and Vegetables,* edited by David Scofield Wilson and Angus Kress Gillespie, 166–87. Knoxville: University of Tennessee Press, 1999.

Tannock, Stuart. "Nostalgia Critique." *Cultural Studies* 9 (1995): 453–64.

Tatham, William. *An Historical and Practical Essay on the Culture and Commerce of Tobacco.* 1800. Reprinted in *William Tatham and the Culture of Tobacco,* by Melvin G. Herndon, 1–330. Coral Gables: University of Miami Press, 1969.

Tiller, Kelly. "Tobacco Buyout Top Ten." Prepared for the University of Kentucky, North Carolina State University, and the University of Tennessee, February 21, 2005. http://agpolicy.org/tobaccobuyout/tobuy/TopTen.pdf (no longer available).

Tobacco Institute. "Kentucky's Tobacco Heritage." Pamphlet. Washington, DC: Tobacco Institute, n.d.

Twelve Southerners. *I'll Take My Stand: The South and the Agrarian Tradition.* 1930. Reprint, Baton Rouge: Louisiana State University Press, 1977.

Umberger, Eugene Baker, Jr. "Tobacco Farming: The Persistence of Tradition." MA thesis, Center for Intercultural and Folk Studies, Western Kentucky University, Bowling Green, 1975.

United States Department of Agriculture. *2007 Census of Agriculture: County Profile, Henry, Kentucky.* National Agriculture Statistics Service, February 2009. http://www.agcensus.usda.gov/Publications/2007/Online_Highlights/County_Profiles/Kentucky/index.asp.

———. *2007 Census of Agriculture: Demographics.* National Agriculture Statistics Service. http://www.agcensus.usda.gov/Publications/2007/Online_Highlights/Fact_Sheets/demographics.pdf. Accessed August 2, 2009.

———. *2007 Census of Agriculture: Kentucky State and County Data.* National Agriculture Statistics Service, February 2009. http://www.agcensus.usda.gov/Publications/2007/Full_Report/Census_by_State/Kentucky/.

———. *2007 Census of Agriculture: North Carolina State and County Data.* National Agriculture Statistics Service, February 2009. http://www.agcensus.usda.gov/Publications/2007/Full_Report/Census_by_State/North_Carolina/index.asp.

———. *2007 Census of Agriculture: United States Summary and State Data.* National Agriculture Statistics Service, February 2009. http://www.agcensus.usda.gov/Publications/2007/Full_Report/index.asp.

———. *Kentucky Agricultural Statistics and Annual Report 2006–2007.* Washington, DC: National Agriculture Statistics Service, 2007.

Van Willigen, John, and Susan C. Eastwood. *Tobacco Culture: Farming Kentucky's Burley Belt.* Lexington: University Press of Kentucky, 1998.

Wagner, Susan. *Cigarette Country: Tobacco in American History and Politics.* New York: Praeger, 1971.

Waldrep, Christopher. *Night Riders: Defending Community in the Black Patch, 1890–1915.* Durham: Duke University Press, 1993.

Walker, Melissa. *Southern Farmers and Their Stories: Memory and Meaning in Oral History*. Lexington: University Press of Kentucky, 2006.
Warren, Robert Penn. *Night Rider.* Boston: Houghton Mifflin, 1939.
White, Hayden. *Metahistory: The Historical Imagination in Nineteenth-Century Europe.* Baltimore: John Hopkins University Press, 1973.
Wright, D. Wynne. "Fields of Cultural Contradictions: Lessons from the Tobacco Patch." *Agriculture and Human Values* 22, no. 4 (2005): 465–77.

Interviews

Armstrong-Cummins, Karen. Interview with John Klee. June 6, 2000. Digital recording. Frankfort: Kentucky Oral History Commission.
Baesler, Alice. Interview with the author. February 1, 2008. Digital recording. Frankfort: Kentucky Oral History Commission.
Bishop, Keenan. Interview with the author. August 2007. Digital recording. Frankfort: Kentucky Oral History Commission.
Bogey, Alvin. Interview with the author. September 16, 2005. Digital recording. Frankfort: Kentucky Oral History Commission.
Bond, Jerry. Interview with the author. September 15, 2005. Digital recording. Frankfort: Kentucky Oral History Commission.
———. Interview with the author. September 16, 2005. Digital recording. Frankfort: Kentucky Oral History Commission.
Bond, Kathleen, and Jerry Bond. Interview with the author. August 21, 2005. Digital recording. Frankfort: Kentucky Oral History Commission.
Carter, Mike. Interview with the author. June 7, 2007. Digital recording. Frankfort: Kentucky Oral History Commission.
———. Interview with the author. August 23, 2007. Digital recording. Frankfort: Kentucky Oral History Commission.
Cline, Bruce. Interview with Lynne David. September 21, 2001. Digital recording. Frankfort: Kentucky Oral History Commission.
Corn, Maurice. Interview with the author. March 22, 2007. Digital recording. Frankfort: Kentucky Oral History Commission.
Crain, W. Ben. Interview with the author. December 17, 2007. Digital recording. Frankfort: Kentucky Oral History Commission.
———. Interview with the author. February 4, 2008. Digital recording. Frankfort: Kentucky Oral History Commission.
———. Interview with John Klee. July 31, 2000. Digital recording. Frankfort: Kentucky Oral History Commission.
Davis, Layton. Interview with Lynne David. October 24, 2001. Digital recording. Frankfort: Kentucky Oral History Commission.
Duncan, George. Interview with the author. November 29, 2007. Digital recording. Frankfort: Kentucky Oral History Commission.

———. Interview with the author. December 5, 2007. Digital recording. Frankfort: Kentucky Oral History Commission.
Evans, Jenny, and Kevan Evans. Interview with the author. September 18, 2007. Digital recording. Frankfort: Kentucky Oral History Commission.
———. Interview with Kara Keeton. September 24, 2006. Digital recording. Frankfort: Kentucky Oral History Commission.
Gallagher, Clarence. Interview with the author. July 20, 2007. Digital recording. Frankfort: Kentucky Oral History Commission.
Glass, Frances, and Willard Glass. Interview with the author. September 17, 2005. Digital recording. Frankfort: Kentucky Oral History Commission.
Grigson, Dan. Interview with the author. September 27, 2007. Digital recording. Frankfort: Kentucky Oral History Commission.
Grigson, Valerie. Interview with the author. September 6, 2007. Digital recording. Frankfort: Kentucky Oral History Commission.
Harkins, Susan. Interview with the author. September 5, 2008. Digital recording. Frankfort: Kentucky Oral History Commission.
Henson, Kathy. Interview with the author. September 12, 2007. Digital recording. Frankfort: Kentucky Oral History Commission.
Henson, Martin. Interview with the author. February 7, 2007. Digital recording. Frankfort: Kentucky Oral History Commission.
———. Interview with the author. April 27, 2008. Digital recording. Frankfort: Kentucky Oral History Commission.
Hinton, Madonna. Interview with the author. March 9, 2007. Digital recording. Frankfort: Kentucky Oral History Commission.
Hopkins, Eddie, and Pam Hopkins. Interview with the author. June 19, 2007. Digital recording. Frankfort: Kentucky Oral History Commission.
Keugel, William "Rod." Interview with Lynne David. October 23, 2001. Digital recording. Frankfort: Kentucky Oral History Commission.
Long, Charlene, and Charles Long. Interview with the author. September 17, 2005. Digital recording. Frankfort: Kentucky Oral History Commission.
McKinney, Danny. Interview with John Klee. June 6, 2000. Digital recording. Frankfort: Kentucky Oral History Commission.
Meredith, Ben. Interview with the author. September 13, 2008. Digital recording. Frankfort: Kentucky Oral History Commission.
Miller, Judy. Interview with John Klee. September 28, 2000. Digital recording. Frankfort: Kentucky Oral History Commission.
Moore, Steve. Interview with the author. November 20, 2007. Digital recording. Frankfort: Kentucky Oral History Commission.
Morse, Donald. Interview with the author. December 17, 2007. Digital recording. Frankfort: Kentucky Oral History Commission.

Palmer, Gary. Interview with the author. December 5, 2008. Digital recording. Frankfort: Kentucky Oral History Commission.
Parsons, Ken. Interview with the author. December 5, 2007. Digital recording. Frankfort: Kentucky Oral History Commission.
Pearce, Robert. Interview with the author. August 24, 2007. Digital recording. Frankfort: Kentucky Oral History Commission.
Perkins, Roger, and Lisa Perkins. Interview with the author. December 17, 2007. Digital recording. Frankfort: Kentucky Oral History Commission.
Perkins, Wilbert, and Patsy Perkins. Interview with the author. January 21, 2008. Digital recording. Frankfort: Kentucky Oral History Commission.
Persley, Tom. Interview with the author. May 5, 2007. Digital recording. Frankfort: Kentucky Oral History Commission.
Quarles, Roger. Interview with the author. January 31, 2008. Digital recording. Frankfort: Kentucky Oral History Commission.
———. Interview with the author. May 20, 2008. Digital recording. Frankfort: Kentucky Oral History Commission.
Rankin, Jerry. Interview with the author. January 18, 2008. Digital recording. Frankfort: Kentucky Oral History Commission.
Roberts, Dale. Interview with the author. December 12, 2007. Digital recording. Frankfort: Kentucky Oral History Commission.
Roberts, Mark. Interview with the author. December 12, 2007. Digital recording. Frankfort: Kentucky Oral History Commission.
Sharp, Phil, and Phyllis Sharp. Interview with the author. March 24, 2007. Digital recording. Frankfort: Kentucky Oral History Commission.
Shell, Jonathan, and G. B. Shell. Interview with the author. March 19, 2007. Digital recording. Frankfort: Kentucky Oral History Commission.
———. Interview with the author. January 22, 2008. Digital recording. Frankfort: Kentucky Oral History Commission.
Snell, Will. Interview with the author. May 22, 2008. Digital recording. Frankfort: Kentucky Oral History Commission.
Souder, Kenny. Interview with John Klee. September 29, 2000. Digital recording. Frankfort: Kentucky Oral History Commission.
Taylor, Carolyn. Interview with the author. July 30, 2005. Digital recording. Frankfort: Kentucky Oral History Commission.
Taylor, David, and Shirlee Taylor. Interview with the author. January 29, 2008. Digital recording. Frankfort: Kentucky Oral History Commission.
Taylor, Diana. Interview with the author. May 23, 2005. Digital recording. Frankfort: Kentucky Oral History Commission.
Taylor, Robert. Interview with the author. July 20, 2005. Digital recording. Frankfort: Kentucky Oral History Commission.
Waits, Marlon. Interview with the author. September 22, 2007. Digital recording. Frankfort: Kentucky Oral History Commission.

Waits, Ray, and Harold Waits. Interview with the author. December 6, 2007. Digital recording. Frankfort: Kentucky Oral History Commission.
Wallace, Dean. Interview with the author. January 28, 2008. Digital recording. Frankfort: Kentucky Oral History Commission.
———. Interview with John Klee. July 13, 2000. Digital recording. Frankfort: Kentucky Oral History Commission.
White, Scott. Interview with John Klee. June 26, 2000. Digital recording. Frankfort: Kentucky Oral History Commission.
Wise, Noel. Interview with the author. April 27, 2007. Digital recording. Frankfort: Kentucky Oral History Commission.

Index

Page numbers in *italics* refer to photographs. Page numbers that appear in **bold** refer to maps. The letter *f* following a page number denotes a figure.

Kentucky Remembered: An Oral History Series

James C. Klotter,
Terry L. Birdwhistell,
and Douglas A. Boyd,
Series Editors

Books in the Series

Conversations with Kentucky Writers
edited by L. Elisabeth Beattie

Conversations with Kentucky Writers II
edited by L. Elisabeth Beattie

Barry Bingham
Barry Bingham

Crawfish Bottom: Recovering a Lost Kentucky Community
Douglas A. Boyd

This Is Home Now: Kentucky's Holocaust Survivors Speak
Arwen Donahue and Rebecca Gayle Howell

Burley: Kentucky Tobacco in a New Century
Ann K. Ferrell

Freedom on the Border: An Oral History of the Civil Rights Movement in Kentucky
Catherine Fosl and Tracy E. K'Meyer

Arab and Jewish Women in Kentucky: Stories of Accommodation and Audacity
Nora Rose Moosnick

Bert Combs the Politician
George W. Robinson

Tobacco Culture: Farming Kentucky's Burley Belt
John van Willigen and Susan C. Eastwood

Food and Everyday Life on Kentucky Family Farms, 1920–1950
John van Willigen and Anne van Willigen

Voices from the Peace Corps: Fifty Years of Kentucky Volunteers
Angene Wilson and Jack Wilson

CPSIA information can be obtained at www.ICGtesting.com
Printed in the USA
BVOW040132200613

323791BV00002B/5/P

9 780813 142333